A Practical Guide to Behavioral Research

A Practical Guide To Behavioral Research

Tools and Techniques

THIRD EDITION

Barbara Sommer
Robert Sommer

New York Oxford
OXFORD UNIVERSITY PRESS
1991

Oxford University Press

Oxford New York Toronto
Delhi Bombay Calcutta Madras Karachi
Petaling Jaya Singapore Hong Kong Tokyo
Nairobi Dar es Salaam Cape Town
Melbourne Auckland

and associated companies in
Berlin Ibadan

Published by Oxford University Press, Inc.,
200 Madison Avenue, New York, New York 10016

Library of Congress Cataloging-in-Publication Data
Sommer, Barbara Baker, 1938–
A practical guide to behavioral research :
tools and techniques /
Barbara Sommer and Robert Sommer. — 3rd ed.
p. cm.
Includes bibliographical references and indexes.
ISBN 0-19-506712-6.—
ISBN 0-19-506713-4
1. Social sciences—Research.
2. Social sciences—Methodology.
3. Psychology—Research.
4. Social sciences—Research—Data processing.
I. Sommer, Robert. II. Title.
H62.S724734 1991 150′.72—dc20 90-47753

9 8 7 6 5

Printed in the United States of America
on acid-free paper

Preface

Since the 1986 edition, important changes have occurred in the behavioral sciences, in the outside world, and in ourselves. A sabbatical leave spent teaching in Estonia, U.S.S.R, made us aware of research taking place in other nations under conditions very different from those we had known and written about. Topics considered important in the West may have little meaning or importance elsewhere, and vice versa. This edition includes more reference to work outside North America. Examples have also been updated. Although we are sentimentally attached to some of the earlier studies (oldies but goodies), we recognize that students appreciate contemporary examples, studies "in their own time," rather than in the time of the instructor and textbook authors.

The book is suitable for students in research methods courses and for nonstudents who wish for a clear guide to conducting behavioral studies. These techniques are too important to be left in the hands of a small group of specialists. The long-run interests of a democratic society will be served by the widest possible understanding and acceptance of behavioral research methods.

Reflecting the heightened interest in qualitative approaches, the observation chapter has been divided into two sections, with the first on more traditional systematic observational techniques, and the second on participant observation and ethnography.

The chapter on library usage has been brought into the computer age. There are new electronic information indexes that students should know about even if they are not immediately available. We attempt to provide some guidance to minimize information overload. The computer section has been rewritten to reflect the widespread availability of PCs and statistical programs. However, we do not want the tail (statistical analysis) to wag the dog (all of the other aspects of research). The computer has not changed the need for properly designed experiments, carefully worded questionnaires, interviews without bias, and reliable observation categories. These issues remain the central foci of the book. There is a new section in Chapter 20 on preparing a critique of a scientific article. This is an important element in the peer review process and in becoming an informed consumer of research studies.

We are indebted to those colleagues and students who read and evaluated the earlier editions, whose comments helped remove ambiguity, jargon, and the occasional error. Ron Goldthwaite, Laurie Irish, Gary Mitchell, Richard Rasor, and Michael Winter made helpful comments.

Carol Van Alstine and Marilee Urban provided valuable clerical help. We are grateful to the literary executor of the late Sir Ronald A. Fisher, to Dr. Frank Yates, and to Longmans Group, Ltd., London, for permission to reprint several tables from their book *Statistical Tables for Biological, Agricultural, and Medical Research* (6th ed., 1974).

Davis, California
April 1990

B.S.
R.S.

Contents

I INTRODUCTION TO RESEARCH

I
Introduction to Research

1 Multimethod Approach

What Is Behavioral Research?

Behavioral research is everybody's business. It is difficult to imagine an occupation in which systematic information about human behavior is not relevant. Forest rangers are increasingly concerned about vandalism, arson, and litter in wilderness areas. NASA wants to know how groups of astronauts will react during space flights to distant planets. Food processors are concerned about the public response to new products. The lives of all of us have been touched, directly or indirectly, by the procedures and results of behavioral research.

There is an important difference between thinking about human behavior and doing research. Research is careful, patient, and methodical inquiry done according to certain rules. It is not simply an exchange of views among friends, colleagues, or experts. Anyone whose job depends on information about what people do or want should know how to obtain that information in a valid and systematic manner. Specifically, they should know how to interview, construct a questionnaire, observe natural behavior, and conduct an experiment.

Even professions whose primary concerns are with machines must take account of human behavior. What is the most efficient way to arrange letters on a typewriter keyboard? What is the best way to arrange instrument dials in a jet aircraft to minimize risks of confusion and error? The answers to such questions will be based, to some degree, on the results of behavioral studies. However, not all questions are behavioral. A comparison of the gas consumption of two types of engines is an engineering problem. Yet at some point that question will have behavioral aspects, perhaps in terms of the noise level of the two engines. Behavioral research cannot answer nonbehavioral questions. Whether adultery is good or evil is a moral and not a behavioral question. The interpretation of a court decision is a legal question. Such questions have behavioral implications, but the researcher must realize what parts of them are behavioral and what parts are not.

This book emphasizes systematic methods for gathering information. Visual inspection of a shopping mall is better than nothing, but a systematic observational study over a period of time will yield information that is more accurate and believable, and probably more useful. The game warden who believes that fluorescent blue or yellow would be more suitable than orange for hunting jackets under conditions of poor visibility could question several hunters on the issue. A better approach would be to do an experiment to determine the appearance of each color under different degrees of illumination. The range of questions that can be examined using the methods of behavioral research is enormous. It is the task of the researcher to set priorities for the questions to be asked, as well as to identify the best methods to be used.

Theory and Research

There are two components to a scientific body of knowledge: empirical research and theory. *Empirical* research involves the measurement of observable events—for example, the effect of a particular drug on reaction time, people's responses to questionnaires, or individual characteristics measured by a personality inventory. Empirical refers to information that is sense-based—what we directly see, hear, touch, smell, or taste. It is demonstrable; that is, it can be shown to other people. Subjective qualities such as feelings and beliefs become empirical when expressed by means of attitude scales, interviews, ratings, or some other measurement procedure.

Theories are logical constructions that explain natural phenomena. They are not in themselves directly observable, but can be supported or refuted by empirical findings. Theory and empirical research are connected by means of *hypotheses*—testable propositions that are logically derived from theories. The "testable" aspect is very important in that scientific hypotheses must be capable of being accepted or rejected. For example, many religious statements are not directly testable in that they can be neither supported nor refuted on the grounds of direct observation. Thus they do not fall within the realm of science. The proposition that "People were placed on earth to rule over other creatures" is not directly testable in an empirical sense.

There is a reciprocal relationship between theory and research. Research shapes a theory by bringing it into accord with the observable world and thereby increasing the theory's explanatory and predictive power. Theories not only link a number of sometimes-diverse observations into coherent wholes, but they also generate the hypotheses that stimulate further research and further refinement of theories. This book is about doing empirical behavioral research.

Behavioral research is conducted in accord with the scientific method. As noted, its subject matter is directly observable, or can be made so through some type of representation (e.g., questionnaire responses). The data, such as the observations or productions, are subject to *consensual validation;* that is, independent observers agree to what is being represented. These data can be repeated or replicated under the same conditions. The condition of repeatability cannot always be met, as in case studies of one-time events—a problem analogous to that of astronomers who carry out scientific studies on unique events. Scientific terms should be precise and clearly defined. Generally, a scientist uses an *operational definition;* that is, defining something by the means used to measure it. For example, hunger might be defined as hours since last eating, or racial attitudes by a score on an attitude scale. Sometimes operational definitions may not be agreed upon—for example, defining masculinity as being able to grow a beard—but they have the advantage of being clear, and thus open to scrutiny.

Validity and Reliability

Validity is the degree to which a procedure measures what it is supposed to measure. Is the operational definition consistent with other ways of identifying and measuring the behavior or characteristic? Defining sociability in terms of the number of interactions that a person has each day would be of questionable validity because it includes people whose occupations or other responsibilities involve many forced encounters. A more valid measure would take into account whether an interaction was more or less voluntary. Number of hours since eating is probably a valid measure of hunger, although hunger is a subjective experience; therefore, the validity of the operational definition rests on its adequacy in meeting the research objectives.

External validity refers to the generalizability of the findings. Do the results extend beyond the immediate setting or situation? Research in natural settings often provides higher external validity than does research from the laboratory. In contrast, laboratory research may be higher on *internal validity,* which means that alternative explanations for behavior can be eliminated. Internal validity refers to the ability to rule out competing explanations through control over the situation. In an experiment measuring the effect of sleep deprivation and reaction time, the researcher can systematically rule out other factors that may influence the reaction time, such as the type of task performed, the instructions given to the subject, and so on.

In addition to being valid, research should also be *reliable.* Reliability refers to the repeatability or replicability of findings. Instruments and procedures should produce the same results when applied to similar people in similar situations, or to the same people on a second occasion. Reliability is an important contribution to validity. However, a study can be reliable, but not be valid. For example, we might propose strength of hand grip as a measure of intelligence. We can measure hand grip with a high degree of reliability, but that does not mean we have a valid measure of intelligence. The measure lacks internal validity.

Why Do Behavioral Research?

The two most common reasons for doing behavioral studies are (1) to obtain answers to pressing questions and (2) to contribute to theories of human behavior. Providing answers to pressing questions is called *applied research.* It is motivated primarily by the need to solve an immediate problem. One type of applied research that is receiving increasing attention is *program evaluation.* This involves determining the effectiveness of a program. Some government agencies now require evaluation to accompany all new projects. This has created the need for researchers who can adapt themselves to the politically charged and hurried world of legislation, government budgets, and bureaucratic jargon. Program evaluation must be done quickly or else it is useless. There is a growing recognition that evaluation, when done properly, can help to guide legislative action.

Investigations designed to answer general long-range questions about human behavior are considered *basic research* and are motivated largely by the researcher's curiosity. In practice, the division between applied and basic studies is far from clear. Most behavioral research arises from a combination of attempts to answer specific questions and the researcher's curiosity. A third category of behavioral studies, which may include your own interests, is *instrumental research.* This includes studies done as training exercises or as part of a job assignment. Many people doing behavioral research for industry, the military, or government agencies work on problems chosen by others. Some students become intrigued with research done as part of a class project and continue work on it afterward, or the instructor may continue and extend a class project to the point where it becomes a publishable paper.

Although it is useful to make a conceptual distinction between basic and applied research, the line between the two is often blurred and indistinct in practice. A researcher may have mixed motives in conducting a study, seeking both to test part of a theory while answering a practical question at the same time. Social psychologist Kurt Lewin (1946) pioneered *action research* as a means by which behavioral scientists could contribute simultaneously to theory and practice. Lewin and his associates used this approach in communities attempting to reduce racial prejudice. The programs to change atti-

BOX 1-1. Four Types of Research Studies

Basic research: Seeks answers to long-range questions. Motivated primarily by curiosity.
Applied research: Seeks practical answers to immediate questions. Goal is to obtain usable information.
Instrumental research: Undertaken as an academic, vocational, or professional requirement. Goal is to demonstrate competence in research.
Action research: Combines the testing of theory with application.

tudes became experiments to be evaluated and revised before they were implemented elsewhere. Action research involves the potential users of the information in doing the research. Lewin believed that the participation of the potential users reduces the gulf between research and application. Other reasons for doing action research are to increase people's involvement in an issue or problem, to develop a constituency for change, to reduce the distance between researchers and the public, to base program changes on sound information (the fruits of research) rather than guesses, to test theories of human behavior in natural settings, and to provide feedback to researchers on the utility of their work.

Not every study needs to be earthshaking. Sometimes a small investigation that answers only a single question is all that is needed. When an occupational therapist in a mental hospital found that many crafts materials were unsuitable for older people with poor eyesight and coordination, she did a brief survey among older women about the problems they had in knitting. Among other things, she found that dull colors made knitting more difficult. She recommended that bright colors were preferable to dull ones for the elderly. This advice was followed in future purchases of craft materials. Although it was not a profound recommendation, it more than justified the small amount of time and effort invested in the survey.

Knowledge Is Connected

An introduction to research techniques would be easier if the methods and disciplines could be kept separate. This is not realistic; there is often a thin, blurred line between one procedure and the next. Constructing a questionnaire will involve interviews. An interview study will be preceded by observations to find out the questions to be asked. Interviews are scored using a method known as *content analysis.* Academic disciplines cannot be kept separate either. A sociologist undertaking participant observation among teenage youth will have more in common with anthropologists than with other sociologists who analyze census records. The psychologist observing children in a nursery may feel more kinship with zoologists than with laboratory-based experimental psychologists. It is important for students interested in human behavior to learn a variety of techniques. This will not only help them choose the best methods for a project but will also be indispensable for understanding relevant studies from other disciplines. To understand why people act as they do, one cannot view their actions solely in psychological terms, economic terms, or historical terms. Knowledge will be more thorough and accurate when it includes ideas and findings from many different sources.

An interdisciplinary approach is essential for research on complex problems. A small study of only one aspect of a problem, such as the relationship between airplane engine types and the amount of noise, might involve only a mechanical engineer and an acoustical consultant, but a larger study on airport noise might include an engineer, architect, physician or public health worker, and a behavioral scientist. Team members must avoid the jargon of individual specialties and develop a shared language. Each single investigation must be integrated into the larger effort and timetable.

Practical Experience

Specific techniques can be learned from a book. You can read the definition of a questionnaire, how it differs from an interview, and its advantages and disadvantages. To acquire skill in using the questionnaire as a research tool requires something more—experience. The difficulty in writing clear questions may not be evident yet. Thus, when you construct your first questionnaire, you are likely to make all the mistakes you have read about. Nor is it obvious how boring it is to spend an hour watching children on the playground for more than one day. Sixty minutes never seemed so long. Another thing that may not be apparent yet is the intellectual thrill of putting together a mass of data you have collected that helps to clarify a complex problem. The emotional side of research, the joy and excitement, as well as the frustration and boredom, are best learned by *doing* research. Like other skills, specific research techniques are learned most readily by following accepted procedures. After some proficiency is acquired, the student can decide when to ignore the rules.

With experience, you will find yourself more attracted to some procedures than to others. Don't worry if one or more techniques turns you off. This is also true of practicing researchers. Some like nothing more than spending their days in a laboratory using sophisticated equipment. These same experimenters may feel uncomfortable conducting observations in the field where nothing is controlled. Other researchers enjoy sending out questionnaires in mail surveys and letting the postal service do the work. Some researchers do not collect original data but rely on secondary sources, such as health statistics and census records, or undertake content analyses of the mass media. Depending upon the problem studied, each of these approaches has the potential to yield theoretically significant or useful information.

Even if you dislike a particular method, it is still important to learn about it so that you can evaluate the work done by others. You may someday find yourself part of an interdisciplinary team whose members employ different methods. If you consider a technique to be dull or difficult, it is useful to reflect upon the reasons, since this will help you define your personal research style. Presumably you will do a better job using methods with which you feel comfortable. Knowledge of your own preferences will help you set priorities for future involvement in research projects.

Multimethod Approach

Each technique for gathering information has its shortcomings. Experimentation is limited by artificiality, observation by unreliability, interviews by interviewer bias, and so on. There is no ideal research technique in the behavioral sciences. The advantages may lie along one dimension, such as economy; the disadvantages along another, such as objectivity. The goal of the researcher is not to find the single best method. For most problems, several procedures will be better than one. Even though each has its limitations, these tend not to be the same limitations. The artificiality of the laboratory can be supplemented by observation, which is high on naturalness but low on reliability; the questionnaire, which can be given to many people quickly, can be supplemented by detailed interviews with a few people to probe more deeply into significant issues. This has been described as the method of *converging operations* (Webb, Campbell, Schwartz, & Sechrest, 1966). A number of different research techniques are applied, each with somewhat different limitations and yielding somewhat different data. Conflicts between the information from different sources will sometimes occur. In one case, the letters received by a director of a veterinary hospital consisted mostly of complaints. On the other hand, interviews with pet owners bringing their animals to this same hospital showed a high level of satisfaction (Sommer & Tyburczy, 1978). There was no basic contradiction between the two sources of information. People dissatisfied with the hospital wrote letters of complaint and went elsewhere, while satisfied customers continued to visit the hosptial. Both the complaints and the interviews contributed to an understanding of how the hospital worked.

The multimethod approach provides flexibility in dealing with obstacles encountered in carrying out a project. Sometimes the most appropriate procedure cannot be used so the researcher must fall back on a combination of other techniques. When experimentation is not possible, the researcher may use a combination of observations and interviews, and if neither of these is possible, the researcher will try simulation (creating an artificial reality in the laboratory), or read autobiographical accounts. Having a variety of methods available, even when they are not all used, provides a flexibility beyond what is possible with a single procedure.

Studying complex issues virtually requires multiple methods, not only for breadth of coverage, but also to allow for a check on the validity of individual methods. To understand how people in Germany were affected by the nuclear accident at Chernobyl in the Soviet Union, Kaminski (1988) combined interviews with content analyses of media. Members of his research team interviewed government officials and experts who had manned special telephone hotlines installed shortly after the accident. They also interviewed people most likely to be affected by the accident, such as pregnant women, mothers with small children, farmers, and greengrocers. They tabulated letters and calls to television and radio stations, and collected a year's issues of the city newspaper to analyze all references to the accident.

Box 1-2 shows two examples of a multimethod approach in very different situations. In the first example, a team of investigators was interested in children's classroom study behavior. No single method seemed capable of capturing all aspects of this phenomenon. The solution was to combine observations with interviews and questionnaires. The second study was an evaluation of a jury system in architectural education. This is a stressful procedure in which a student's work is publicly evaluated by a panel of instructors and outside professionals. The combination of observations, interviews, diaries, and questionnaires, seemed preferable to any single method used alone.

The major advantage of the multimethod approach is not the quantity of data collected, but the diversity of data collected on a single topic and the opportunities for comparison that this diversity affords (Brewer and Hunter, 1989). To be of most value, the data obtained from the different methods should be independent. When more than a single procedure is used in a setting, there is the risk that the first procedure may affect the results of subsequent ones. The effect of the research upon the participants and the data collected is termed *reactivity.*

One method for reducing reactivity is to partition the sample and use different procedures in each sub-sample. In a survey using both a mail questionnaire and interviews, the researcher could divide the potential subjects randomly into two different groups, using interviews with one group and mail questionnaires with the second. Another way to minimize reactivity among different procedures is to carefully plan the sequence in which they are used, starting with the least and going to the most reactive procedures. As an example, in combining observations and interviews, the rule would be to look first and ask questions later. Since the interview is generally a more intrusive and reactive procedure than natural observation, it makes good sense to complete the observations before beginning the interviews. Mailed questionnaires are generally less reactive than personal interviews, so if you were combining these two approaches, the mail survey should precede the personal interviews.

BOX 1-2. Two Examples of Multimethod Research

A. Children's Study Behavior

To understand how children study in the classroom, Lyman, Stock, and Kulhavy (1989) used the following methods:

A. Structured observations by two observers, both experienced teachers, in classrooms for grades 1–6. The data consisted of both written and verbal descriptions of study behaviors.

B. Based on the observations, a questionnaire was constructed and revised. The revised version was administered to 803 elementary school students, divided about equally across second, fourth, and sixth grades.

C. Ratings of some of the key items were obtained from 18 elementary school teachers.

B. Evaluation of Architectural Juries

Architectural juries can often be very stressful to students whose work is evaluated. Students whose work has been publicly criticized make statements like, "My project was ripped to shreds" or "We were massacred." Surprisingly, there has been very little systematic research on the use of this form of public evaluation. To gain a full picture of how juries operated and their effects upon the participants, Anthony (1987) combined the following techniques;

A. Systematic observations of 9 jury sessions, some videotaped, for a total of 130 student presentations.

B. Interviews with 43 students immediately after receiving criticism from the jury, another 40 students on an ongoing basis throughout the term, and interviews with 19 faculty who served on the juries.

C. Diaries from 27 students, with special emphasis on such items as anxiety, eating and sleeping patterns before the jury session and so on.

D. Questionnaires completed by 151 architectural students, 81 students from other fields, and 34 practicing architects.

National Differences

Behavioral research is intimately connected to place and time. The choice of topic, method, and even the wording of the questions must be suited to local conditions. A printed questionnaire will not be useful in a developing nation with a high rate of illiteracy. Observational techniques will not be appropriate in an authoritarian nation whose citizens worry about being spied upon.

Experimental psychologists doing research in developing nations will probably not be able to afford expensive equipment and facilities. Access to sophisticated computers will be difficult in certain parts of the world. During our 4-month teaching visit in the Soviet Union, copy machines were unavailable. This made it very difficult for us to do any questionnaire research. Another American scholar teaching in the Soviet Union noted the shortage of laboratory equipment:

> After class I talked with a student about an experiment that required a minor piece of equipment that could be easily obtained in any modest-sized city in the United States. He looked at me with great surprise and said, "But I don't have this piece of equipment" and considered the prospects of building one. The absurdity of that would be like replacing the letter X on your typewriter. For Westerners it is simply a matter of contacting a supplier and finding an appropriate key, but the prospects of tooling a separate letter would pose a monumental problem. Needless to say, the research was stymied and would never be done by this student.

There are also important philosophical differences between nations in their attitudes toward behavioral research. Europe has a long tradition of qualitative research in the social sciences. For example, researchers in Germany perceive an overemphasis on quantification and hypothesis-testing among American researchers. Researchers in France are likely to pay more attention to small groups and larger social units than to individuals (Jodelet, 1987). National priorities also affect the choice of research topics. Japan's geographical location has been responsible for a particularly high number of natural disasters, including floods, fires, and earthquakes. This has stimulated a considerable amount of disaster research.

Deciding on Methods

There are dozens of methods available to the behavioral researcher. Not all of these will be equally useful. Four techniques—observation, experiment, questionnaire, and interview—account for more than nine-tenths of the articles in social science journals. Some rules of thumb for selecting among these methods are presented in Box 1-3. Observation is well suited for discovering what people do in public. For private behavior, the personal diary is more appropriate. The experiment is an immensely powerful tool for deciding between alternative explanations of a phenomenon. It is less useful, however, for studying natural behavior or opinions. With opinions and attitudes, the questionnaire and interview are very efficient. Standardized tests are used to assess mental abilities.

In making a decision about methods, the problem comes first. What questions must be answered? The next issue involves the time and resources available. There is no point in planning an elaborate survey if there is no one to carry it out, or to prepare an observational study of a prison whose warden will not permit observers inside the walls. You may have to settle for interviews with ex-prisoners and a reading of prison diaries. Conditions are generally less than ideal in behavioral research. The subject's time is limited, certain techniques are unfeasible because of the setting or ethical considerations, and the researcher's time and budget are finite.

BOX 1-3. Choosing Among Research Techniques

Problem	*Approach*	*Research technique*	*Chapter*
To obtain reliable information under controlled conditions	Test people in a laboratory	Laboratory experiment, simulation	6, 7
To find out how people behave in public	Watch them	Natural observation	4
To find out how people behave in private	Ask them to keep diaries	Personal documents	12
To learn what people think	Ask them	Interview, questionnaire, attitude scale	8, 9, 10
To find out where people go	Chart their movements	Behavioral mapping, trace measures	5
To identify personality traits or assess mental abilities	Administer a standardized test	Psychological testing	15
To identify trends in verbal material	Systematic tabulation	Content analysis	11
To understand an unusual event	Detailed and lengthy investigation	Case study	13

Researchers interested in human stress often encounter ethical problems when they conduct laboratory studies. An alternative is to seek out situations such as a hospital emergency room where stress already exists. A university during final exams would be a good source of tense individuals. Air traffic controllers represent a group for whom stress is an occupational hazard. The multimethod approach provides the option of observing stress as it occurs as well as laboratory studies with animals or human beings. The research team using several linked procedures is not putting all of its eggs into a single basket. One method can be used first, followed by another, than a third method. Information developed at each stage guides in the selection of the next procedure.

A multimethod approach is also useful in dealing with the unforeseen circumstances that arise in field research. Beginning researchers are likely to be discouraged when they go out to a shopping center and there is no one around to interview because it is too cold, windy, or because it is just an inactive period. One solution is to use such occasions to practice other data-gathering techniques. If there are no customers in the shopping center, then this may be a good opportunity to interview store employees or examine wear on carpets or linoleum, cigarette burns, graffiti, or other residues that will provide clues to usage. The experienced researcher always has several methods in reserve for occasions when the primary method cannot be employed. Making good use of unforeseen circumstances will increase the likelihood of the accidental discovery of new relationships.

Sometimes a researcher is fortunate enough to be able to study a situation or group of people over time. For example, a research team might study children of divorced parents as they grow to adulthood and have children of their own. Research over time is termed *longitudinal,* and is expensive in time and personnel. Although researchers recognize the value of longitudinal studies, in practice few are able to do them. As an alternative, they undertake *cross-sectional* studies in which different groups of people are compared on a single occasion. Instead of studying the same group of children over many years, researchers might compare different groups of 5-, 10-, and 15-year olds.

Summary

Theory and research are two components of scientific knowledge. Empirical research involves systematic measurement of observable events. A theory is a logical construction that explains natural phenomena. Hypotheses, which are testable propositions, connect theory and research. Agreement among independent observers (consensual validation) and defining something by how it is measured (operational definition) are important components of the scientific method.

Validity is the degree to which a procedure measures what it is supposed to measure. External validity refers to the generalizability of the findings. Reliability refers to their repeatability or replicability.

Basic research aims at answering long-range questions. Applied research tries to provide solutions for immediate problems. Instrumental research is undertaken as either a training exercise or a job assignment. Action research combines the testing of theory with application.

An interdisciplinary approach is essential for research on complex social issues. For most problems, several procedures will be better than one. Even though each method has its limitations, these tend not to be the same limitations. In choosing between alternative methods, the problem comes first; the next issues to be considered are the time and resources available. Longitudinal studies extend across time while cross-sectional research investigates different groups or events on a single occasion.

References

Anthony, K. H. (1987). Private reactions to public criticism. *Journal of Architectural Education, 40,* 2–11.

Brewer, J. & Hunter, A. (1989). *Multimethod research.* Newbury Park, CA: Sage.
Jodelet, D. (1987). The study of people-environment relations in France. In D. Stokols & I. Altman (Eds.). *Handbook of environmental psychology* (pp. 1171–1193). New York: Wiley.
Kaminski, G. (1988). *Is the development of a psychological ecology useful and possible?* (Report No. 28). Tuebingen, West Germany: Psychological Institute, University of Tuebingen.
Lewin, K. (1946). Action research and minority problems. *Journal of Social Issues, 2,* 34–36.
Lyman, S. L., Stock, W. A., & Kulhavy, R. W. (1989, August). *Critical incidents of study behavior.* Paper presented at the annual convention of the American Psychological Association, New Orleans.
Sommer, R. & Tyburczy, J. (1978, March). Waiting rooms: A study in user perception. *California Veterinarian, 32,* 21–23.
Webb, E. J., Campbell, D. T., Schwartz, R. D. & Sechrest, L. (1966). *Unobtrusive measures: Nonreactive research in the social sciences.* Chicago: Rand McNally.

Further Reading

Brewer, J., & Hunter, A. (1989). *Multimethod research.* Newbury Park, CA: Sage.
Brinberg, D., & McGrath, J. E. (1985). *Validity and the research process.* Beverly Hills, CA: Sage.

2 Ethics in Behavioral Research

Most universities and scientific institutes require researchers to fill out special forms before they can conduct research with human beings. These forms are sent to an Institutional Review Board (IRB), which looks them over to decide if the project should be allowed to proceed. The procedure for ethical review on our campus typifies that of most universities. Investigators submit one of several forms depending upon the type of research undertaken. The following categories of studies are usually exempt from reviews (i.e., do *not* require a formal review):

1. Studies conducted in established educational settings involving formal educational practices, for example, evaluation of different teaching methods and curricula.

3. Research using educational tests when the information is recorded in such a manner that respondents cannot be identified.

3. Surveys or interviews in which the respondents cannot be identified, and when the responses do not place the respondents at risk.

4. Observation of people in public locations, except when the observations include identifying information, might place the individual at risk, or deal with problematic behaviors.

5. Research using existing records or public documents, or if the information is recorded by the investigator in such a manner that people cannot be identified.

Researchers who believe that their studies meet the preceding conditions submit a "Statement of Exemption" to the Institutional Review Board, which is responsible for determining that the conditions for exemption have been met.

Many student research projects fit into the exempt category. This is particularly true in observational and questionnaire studies where the respondents' names are not obtained. For nonexempt studies whose methods are scrutinized by the Institutional Review Board, a more detailed research approval form is used. Box 2-1 shows a completed Human Subjects Approval Form for a student project. All relevant questionnaires and rating scales should be attached to the approval form when it is submitted.

Box 2-2 shows the consent form that subjects were asked to sign for the experiment. Note that the researcher did not mention music on this form, where the study is described in general terms as "learning strategies in university students." It was hoped that omitting mention of music would reduce distortion of the subject's response. Rather than the subjects focusing directly on the music, it would be considered one of the many unusual things that happened during the experiment. At the conclusion of the session, the subject was given a full account of the goals and hypotheses of the study.

BOX 2-1. Example of a Completed Human Subjects Approval Form

1. *Investigator:* Raphael Moore; Faculty Sponsor: R. Sommer.
2. *Title of Research:* Influence of Music on Verbal Retention.
3. *Sponsor Name and Address:* President's Undergraduate Fellowship.
4. *Duration of Study:* Start January 1989; conclude July 1989.
5. *Location:* Social Psychology Laboratory, Young Hall.
6. *Subjects:* 70 undergraduate students.
7. *Contact Method:* Notice on Department bulletin board, in-class announcement.
8. *Procedures:* Survey, questionnaires (drafts attached).
9. *Purpose of Study:* To determine whether a constant rhythm during learning or recall improves retention of nonsense syllables.
10. *Procedure:* Subjects will be randomly divided into four groups: A. sound during memorization; B. sound during recall; C. sound during both memorization and recall; and D. no sound. Students will be tested in groups. Each student will have a headset connected to a master control tape unit. The session will begin with recorded instructions. After this a slide with a list of words will be shown. Those in Groups A and C will hear music as the words are being shown. During recall Groups B and C will hear music.
11. *Risks and Benefits:* The procedure poses no potential risk to the subjects. After the session is finished, students will receive a description of the objectives and procedures, and can request a copy of the data analysis. The experience should be of educational value to the students.
12. *Informed Consent:* See attachment in Box 2-2.

BOX 2-2. Example of a Completed Consent Form

CONSENT TO PARTICIPATE IN RESEARCH STUDY

TITLE OF STUDY: Learning Strategies in University Students

INVESTIGATOR: Raphael S. Moore

PURPOSE

You are being asked to participate in a research study investigating strategies of learning.

PROCEDURES

A short ability test, measuring your knowledge of words, will first be conducted. You will then follow the instructions heard through your headset, which will include the task of learning words seen on the screen in front of you.

RISKS & BENEFITS

The method of research creates no potential risk to you as a subject. It is a minimal risk study, where the risk–benefit ratio leans heavily toward your benefit. Other than the extra credit you might receive from your psychology class, you may also request printed matter explaining your role in the study. This will include a description of the hypotheses, and final analysis of the data.

CONFIDENTIALITY

Ability test data will be computer coded, and used for analysis only. Original information will be destroyed.

RIGHT TO REFUSE OR WITHDRAW

You may refuse to participate. You may change your mind about being in the study and quit after the study has started.

QUESTIONS

If you have any questions, please feel free to ask. If at a later time you have any additional questions, the principal investigator can be reached at 555-8228.

You will be given a copy of the Experimental Subject's Bill of Rights. Also, you may have a copy of this form to keep.

Date	Signature of participant
Date	Signature of investigator

Researchers who study vulnerable populations such as children, the elderly, hospital patients, or jail inmates have special responsibilities in terms of protecting human subjects. Questions can be raised about the legitimacy of informed consent for people in dire or difficult circumstances. These ethical problems are sometimes discussed at professional meetings and in the technical literature. The "Further Readings" section at the end of this chapter lists several good sources. The research community has relied most heavily upon self-regulation in the form of standards developed by professional organizations and institutional review boards that screen proposed research proposals.

The need for these procedures first became apparent in medical research, where people might be exposed to dangerous drugs or radioactive materials. For legal as well as ethical reasons, procedures were needed to ensure (1) that the research plan was reviewed by a committee of competent scientists, (2) the rights and welfare of the subjects were protected, and (3) the subjects were adequately informed of the risks and benefits involved. It was not long before federal granting agencies extended these requirements to behavioral research. Several studies had gained notoriety because of the unorthodox procedures and/or the risks to unknowing participants.

Sign-up board outside Psychology Department office. Announcements describe each experiment and advise students of their rights and responsibilities before they volunteer.

Need for Outside Review

Asking an outside group to make a formal assessment of the risks of a proposed study is always desirable. A researcher can become so involved in a project that some ethical problems are overlooked. The researcher may not be aware of all the risks involved. Even experts can disagree about the likelihood or amount of risk to participants. Some research carries hazards for society or for the researcher. The issue is whether the potential benefits are worth the risk, and the people subjected to the risk are informed of its nature and magnitude. Full disclosure can pose special problems for the behavioral researcher. To tell people the full nature of a study *in advance* can alter their actions. A common solution is to withhold information during the session and debrief the participants afterward. Some researchers will tell people that they are taking part in a psychological experiment that will not involve pain or any physical danger, and ask them for permission not to tell them about the experiment until afterward in order to obtain unbiased responses. Immediately following the session, each participant is given a printed sheet describing the purposes and methods of the study, and the experimenter is present to answer any questions.

Confidentiality and Anonymity

Two methods used by researchers to protect participants in behavioral studies are confidentiality and anonymity. *Confidentiality* means that the respondent's identity is known to the investigator but protected from public exposure. The researcher keeps any identifying information out of published reports. Confidentiality is particularly important when people's statements or actions would cause them some embarrassment if they became known. It is best for the researcher to be on the safe side regarding confidentiality, since it is difficult to predict how people's answers might be interpreted or used by others. Instead of identifying organizations or cities by name, a general description, such as "a middle-sized industrial city in the Northeast" or a fictitious name such as Yankee City or Worktown, is used instead.

The best way to ensure that the people you have interviewed or observed will not be embarrassed by your data is to remove identifying information, such as names and addresses, as soon as the data are tabulated. Some researchers use a special code at the time of the interview so that no names or other identifying information are in the researcher's files. The exact methods used to protect your respondents will vary according to the situation, but it is important to realize that behavioral researchers do not have the special right to confidential communication that the courts grant to physicians, lawyers, and the clergy.

Anonymity means that the researcher does not know the identity of the participants in the study. The best protection for people in an observational study is anonymity. It makes more sense to observe people in a public park or courtyard without knowing their names than to ask each person to sign a written consent form. Anonymity is also a useful safeguard in questionnaire and public opinion research. Most questionnaire studies do not ask people to sign their names. Often this is emphasized by telling respondents *not* to sign their names.

The ethical dilemmas in behavioral research should not be exaggerated or overdramatized. Most surveys, observational studies, and laboratory experiments pose little or no risk to the participants. Students asked to learn lists of words could be told the exact nature of the procedure, what is going to be done with the results, and so on. People whose movements are charted throughout the day would be told of the potential inconvenience of having an observer following them around. The researcher may decide to reserve information about how the results are to be used (e.g., comparing the time younger and older people spend sleeping) until the observations are concluded, but this withholding does not increase the risks to the participants, who know exactly what the researcher will be doing.

Use of Deception

Among the most difficult ethical decisions facing the behavioral researcher is whether or not to use deception. Deception can range from relatively minor omissions, such as not telling people the full story of what you are doing, to outright falsehood about your identity and the nature of the study. To deceive is to deliberately mislead others. The issue is most relevant in experimentation where personal knowledge of the purposes might change their behavior. However, there is no guarantee that deception will succeed. Rosenthal (1976) found that many participants in experiments were aware of the purposes even when the experimenter attempted to keep the purposes hidden. Deception may have a negative effect on the participant's attitudes toward behavioral research. People dislike being lied to and research subjects are no exception. Finally, there is the negative effect on the researcher when forced to lie to other people. It can produce cynicism and distance from the people you are studying.

With full consideration of the ethical and practical problems in using deception, many researchers find instances where they feel it is warranted. *Impersonation,* or acting as someone other than oneself, has been found useful in understanding life in mental and penal institutions. Researchers had themselves admitted as patients or inmates to see the institution from the inside. To the degree that others are aware of the impersonation, the validity of the observer's experience may be reduced. Consumer researchers may impersonate consumers in order to investigate misleading marketing practices. In this case one deception is used to study another. Social psychologists employ impersonation to investigate discrimination. Researchers of different races pose as prospective apartment renters or house buyers and visit real estate agencies to inquire about the types of housing available. This is probably a more valid method than asking realtors if they discriminate among prospective tenants or homeowners on the basis of race.

Our personal views about deception have been shaped by our experiences. We feel that researchers should try to avoid deception as much as possible. The majority of complaints received by IRBs concern deception (Keith-Spiegel, & Koocher, 1985). Even when it appears that it might be advantageous to mislead others, there are alternative ways to obtain the information without telling lies. Ask yourself whether it is absolutely necessary to deceive other people or whether the same information could be obtained with full disclosure or through some sort of simulation that people know has been staged. Regarding conditions in jail and mental hospitals, where some researchers have used impersonation, others have created artificial prisons and jails in which volunteer subjects spent various periods of time in "captivity." As described in Chapter 7, important aspects of confinement have been studied in these simulation environments without the use of deception.

Psychologists' Code of Ethics

Professional organizations in the social sciences have developed guidelines for the ethical conduct of research with human participants. The first version of the ethical code for members of the American Psychological Association was developed in 1953, and has been revised several times on the basis of the experiences of researchers. The code for psychologists shown in Box 2-3 is accompanied by more detailed descriptions of ethical conduct for each of the 10 principles.

In carrying out your own studies, be sure to find out the procedures within your own organization, agency, or school for protecting research subjects. If you are at a university or research institution, you will probably fill out a form, such as that described at the beginning of this chapter, that will be reviewed by an official committee. *This may involve a delay of several weeks before you can begin your research.* You should plan for this delay beforehand. Don't wait until the last minute to obtain approval from an ethics committee. Filling out these forms is a valuable learning experience. You will have to answer questions that you may not have considered: How much should be told to the participants beforehand? How can the confidentiality of the completed questionnaires be maintained? Will the researcher specify the name of the institution being observed? What if the observer should observe illegal behavior? Answering these questions is a good opportunity to foresee potential problems and develop ways of handling them before they arise.

Certain schools and agencies do not require these questionnaires to be filled out. The number of such institutions is becoming smaller each year as

BOX 2-3. Ethical Principles of the APA

Preamble

Psychologists respect the dignity and worth of the individual and strive for the preservation and protection of fundamental human rights. They are committed to increasing knowledge of human behavior and of people's understanding of themselves and others and to the utilization of such knowledge for the promotion of human welfare. While pursuing these objectives, they make every effort to protect the welfare of those who seek their services and of the research participants that may be the object of study. They use their skills only for purposes consistent with these values and do not knowingly permit their misuse by others. While demanding for themselves freedom of inquiry and communication, psychologists accept the responsibility this freedom requires: competence, objectivity in the application of skills, and concern for the best interests of clients, colleagues, students, research participants, and society. In the pursuit of these ideals, psychologists subscribe to principles in the following areas: 1. Responsibility, 2. Competence, 3. Moral and Legal Standards, 4. Public Statements, 5. Confidentiality, 6. Welfare of the Consumer, 7. Professional Relationships, 8. Assessment Techniques, 9. Research with Human Participants, and 10. Care and Use of Animals.

Acceptance of membership in the American Psychological Association commits the member to adherence to these principles.

Psychologists cooperate with duly constituted committees of the American Psychological Association, in particular, the Committee on Scientific and Professional Ethics and Conduct, by responding to inquiries promptly and completely. Members also respond promptly and completely to inquiries from duly constituted state association ethics committees and professional standards review committees.

Principle 1: Responsibility

In providing services, psychologists maintain the highest standards of their profession. They accept responsibility for the consequences of their acts and make every effort to ensure that their services are used appropriately.

Principle 2: Competence

The maintenance of high standards of competence is a responsibility shared by all psychologists in the interest of the public and the profession as a whole. Psychologists recognize the boundaries of their competence and the limitations of their techniques. They only provide services and only use techniques for which they are qualified by training and experience. In those areas in which recognized standards do not yet exist, psychologists take whatever precautions are necessary to protect the welfare of their clients. They maintain knowledge of current scientific and professional information related to the services they render.

Principle 3: Moral and Legal Standards

Psychologists' moral and ethical standards of behavior are a personal matter to the same degree as they are for any other citizen, except as these may compromise the fulfillment of their professional responsibilities or reduce the public trust in psychology and psychologists. Regarding their own behavior, psychologists are sensitive to prevailing community standards and to the possible impact that conformity to or deviation from these standards may have upon the quality of their performance as psychologists. Psychologists are also aware of the possible impact of their public behavior upon the ability of colleagues to perform their professional duties.

Principle 4: Public Statements

Public statements, announcements of services, advertising, and promotional activities of psychologists serve the purpose of helping the public make informed judgments and choices. Psychologists represent accurately and objectively their professional qualifications, affiliations, and functions, as well as those of the institutions or organizations with which they or the statements may be associated. In public statements providing psychological information or professional opinions or providing information about the availability of psychological products, publications, and services, psychologists base their statements on scientifically acceptable psychological findings and techniques with full recognition of the limits and uncertainties of such evidence.

Principle 5: Confidentiality

Psychologists have a primary obligation to respect the confidentiality of information obtained from persons in the course of their work as psychologists. They reveal such information to others only with the consent of the person or the person's legal representative, except in those unusual circumstances in which not to do so would result in clear danger to the person or to others. Where appropriate, psychologists inform their clients of the legal limits of confidentiality.

Principle 6: Welfare of the Consumer

Psychologists respect the integrity and protect the welfare of the people and groups with whom they work. When conflicts of interest arise between clients and psychologists' employing institutions, psychologists clarify the nature and direction of their loyalties and responsibilities and keep all parties informed of their commitments. Psychologists fully inform consumers as to the purpose and nature of an evaluative, treatment, educational, or training procedure, and they freely acknowledge that clients, students, or participants in research have freedom of choice with regard to participation.

Principle 7: Professional Relationships

Psychologists act with due regard for the needs, special competencies, and obligations of their colleagues in psychology and other professions. They respect the prerogatives and obligations of the institutions or organizations with which these other colleagues are associated.

Principle 8: Assessment Techniques

In the development, publication, and utilization of psychological assessment techniques, psycholgists make every effort to promote the welfare and best interests of the client. They guard against the misuse of assessment results. They respect the client's right to know the results, the interpretations made, and the bases for their conclusions and recommendations. Psychologists make every effort to maintain the security of tests and other assessment techniques within limits of legal mandates. They strive to ensure the appropriate use of assessment techniques by others.

Principle 9: Research with Human Participants

The decision to undertake research rests upon a considered judgment by the individual psychologist about how best to contribute to psychological science and human welfare. Having made the decision to conduct research, the psychologist considers alternative directions in which research energies and resources might be invested. On the basis of this consideration, the psychologist carries out the investigation with respect and concern for the dignity and welfare of the people who participate and with cognizance of federal and state regulation and professional standards governing the conduct of research with human participants.

Principle 10: Care and Use of Animals

An investigator of animal behavior strives to advance understanding of basic behavioral principles and/or to contribute to the improvement of human health and welfare. In seeking these ends, the investigator ensures the welfare of animals and treats them humanely. Laws and regulations notwithstanding, an animal's immediate protection depends upon the scientist's own conscience.

the need for formal review and careful monitoring of human research becomes more widespread. Research done as part of class projects may be exempt from some of the more cumbersome procedures, providing the instructor takes responsibility and there is no risk to the participants. Even if your specific project does not require you to fill out the human subjects questionnaire, the responsibility for maintaining ethical standards remains. You will have to develop your own procedures for protecting participants from inconvenience or risk.

Anthropologists' and Sociologists' Code of Ethics

Anthropologists who conduct their research in other nations and cultures face special problems. They must be concerned about their relationship with host governments and local people. Their actions reflect on their own government. The statement of ethics adopted by the American Anthropological Association describes the anthropologist's responsibilities in six major areas:

1. To the subjects of the study, who are the anthropologists' paramount responsibility. Their welfare, dignity, and privacy come first.
2. The public is owed a commitment to honesty and openness, and a full discussion of the meaning and implication of the research.
3. To the profession of anthropology. Researchers should maintain a level of integrity and helpfulness in the field so that by their behavior and example they will not jeopardize future research there.
4. To students. The researcher must alert students to the ethical problems in research and discourage them from participating in projects employing questionable ethical standards.
5. To the sponsors. Researchers must be honest about their qualifications, capabilities, and goals, avoiding any promises that violate professional ethics. The sponsor must be aware of the researcher's commitment to full disclosure of sources of funds, personnel, and the dissemination of research results.
6. To one's own government and to host governments. Before beginning the study, the researcher must make sure that there will be no governmental interference that compromises the ethical standards of the research.

The American Sociological Association has a code of ethics that is similar in most respects to that of the psychologists and anthropologists but places more emphasis on the proper use of research findings. The code obliges the researcher to report all sources of financial support and any special relationship to the sponsor that might affect the interpretation of the findings.

Reactions to Your Report

Problems of handling your results are often as severe as protection of your respondents. What happens if you find out that someone is not doing the job properly or that employees are dissatisfied with their working conditions? Such conclusions have economic and political implications for others and for you. Don't expect to be regarded as a hero because you have identified problems. Many administrators do not want to hear about dissatisfaction among their employees. You are very likely to encounter immediate and heated resistance. This can take the form of a denial of the severity and extent of the difficulties, sarcastic comments about the impracticality of your recommendations, and the observation that your conclusions are trivial and hardly worth mentioning—What was the point of your study anyway? A critic will seize on some small item, such as a statistical error or misprint, to dismiss the entire report. Even when 98 percent of your findings and recommendations are complimentary, obvious, and harmless, some readers will condemn the entire report on the basis of the other 2 percent that they find objectionable. Don't let this discourage you. Getting people to change long-established procedures is a delicate, difficult, and time-consuming matter. You should try to anticipate the reactions that your report is likely to produce.

Some people will convert your report into an opportunity to criticize an individual or policy that they have disliked for years. These "friends" may severely damage your report. Try to keep the integrity of your findings above local squabbles. The best impression you can create is one of a reasonable and open-minded individual who is interested in what is happening and is not overly dependent on factions within the organization. The knowledge that you are connected with an outside agency that has a tradition for impartial research can help convince others of your commitment to objective inquiry.

One solution to the problem of tentative results, which does not deal with the long-range implications, is for researchers to refrain from presenting results in any public forum until they are definitive. In practical terms, however, legislators and others making policy decisions want to have the best information available even if it is incomplete.

BOX 2-4. Conflict Over Release of Results

A scientist-physician and a public interest physician disagreed publicly over the issue of releasing research results. The scientist, Dr. Savitri Ramcharan, came to Washington, D.C., to present the preliminary findings of her study. These findings showed a link between oral contraceptives and increased rates of cervical cancer. However, Dr. Ramcharan believes it is still too early to become alarmed. She had not, in fact, notified the estimated 15,000 women in her sample of the preliminary results.

The public interest physician—Dr. Sidney Wolf, director of Ralph Nader's Health Research Group—disagreed. Wolf believed that the tentative findings are sufficiently important to alert women and physicians everywhere to the risks of taking the pill.

The basic conflict was over the release of preliminary research results. Should data be made public before the findings are conclusive? "It should always be made public," Dr. Wolf declared. "If the data are good—and hers are—they state that there is an increased amount of cancer among the group of women taking the pill, it is the worse form of paternalism to withhold this information." On the other hand, to release the preliminary results might jeopardize the study itself by provoking an exodus from the pill as well as causing undue alarm if the preliminary results do not hold up. ("A quarrel," 1978)

Summary

Most universities and research institutions have committees to review the ethical aspects of behavioral studies. Two major concerns are protecting the rights and welfare of human subjects and seeing that the subjects are adequately informed about the risks and benefits of their participation. Professional organizations of social scientists have developed guidelines for the ethical conduct of behavioral research.

Two methods used to protect participants in behavioral studies are confidentiality and anonymity. Confidentiality means that the respondent's identity, while known to the investigator, is protected from public exposure. Anonymity means that the researcher does not know the identity of the participants in the study. Use of deception in research should be avoided whenever possible.

Limitations: Ethical guidelines phrased in general terms are often difficult to apply in specific instances. It is also difficult to specify in advance all the ethical problems that might arise in the course of the study. Committees charged with protecting the rights and welfare of research subjects may be inflexible in interpreting rules and policies and thus discourage valuable and necessary research. On the other hand, committees may be unfamiliar with the risks in frontier areas of research and allow studies which expose people to unwarranted risks.

References

A quarrel over releasing pill data. (1978, May 23). *Sacramento Bee,* p. A4.

American Anthropological Association. (1973). *Professional ethics: Statements and procedures.* Washington, DC: Author.

American Psychological Association. (1990). Ethical principles of Psychologists. *American Psychologist, 45,* 390–395.

American Sociological Association. (1984). *Code of ethics.* Washington, DC: Author.

Keith-Spiegel, P., & Koocher, G. P. (1985). *Ethics in psychology: professional standards and cases.* New York: Random House.

Rosenthal, R. (1976). *Experimenter effects in behavioral research.* New York: Irvington.

Further Reading

American Psychological Association. (1987). *Casebook on ethical principles of psychologists.* Washington, DC: Author.

Bowd, A. D. (1980). Ethics and animal experimentation. *American Psychologist, 35,* 224–225.

Holmes, D. S. (1976). Debriefing after psychological experiments. *American Psychologist, 32,* 858–875.

Keith-Spiegel, P., & Koocher, G. P. (1985). *Ethics in psychology: Professional standards and cases.* New York: Random House.

Kimmel, A. J. (1988). *Ethics and values in applied social research.* Newbury Park, CA: Sage Publications.

Leak, G. K. (1981). Student perception of coercion and value from participation in psychological research. *Teaching of Psychology, 8,* 147–149.

Sieber, J. (Ed.) (1982). *The ethics of social research* (Vols. 1 & 2). New York: Springer-Verlag.

Steininger, M., Newell, J. D., & Garcia, L. T. (1984). *Ethical issues in psychology.* Homewood IL: Dorsey.

Swan, L. A. (1979). Research and experimentation in prisons. *Journal of Black Psychology, 6,* 47–51.

3 How to Do a Literature Review

An important part of a researcher's task is to learn what has been done before. Finding out about previous research is called *reviewing the literature.* Not only will you learn additional information and facts about your area of interest, but you will probably come across new ideas in the search. Don't rush. It is likely that you will come across ideas and methods that will improve your project. It is also exciting to encounter others with a shared interest in your topic. There is not one source of information on a topic, but many.

Source Credibility

There are various places where behavioral research is mentioned. A study may be described on television or in the newspapers, in a secondary source such as a textbook or review article, or it can be seen in its original form in a scientific journal. You should not judge a study based strictly on the source, but there are some guidelines for evaluating sources in terms of probable (but not certain) accuracy, detail, and credibility.

Mass Media

The popular media such as newspapers, television, and magazines, tend to focus on topics or portions of studies that are controversial or sensational. News reports do not present the context of research in terms of previous studies and they frequently omit salient aspects of the methods and findings. This makes it risky to rely on their descriptions of a research study. You should check the original source, which can probably be located with the help of a reference librarian, or the methods described in the following sections.

Secondary Sources

You may also come across mention of a study in a textbook or a review paper. Typically, the description is brief and provides only a general idea of the topic without much information about methods or findings. If you wish to know the details, you need to find the original paper. Most secondary sources contain reference information, such as a journal citation, that allows you to locate the original paper, which is the *primary* source of the information.

Scientific Journal Articles

When an article is published in a scientific journal, you can generally assume it has passed through a *peer review process* (i.e., it has been read and evaluated favorably by researchers familiar with the topic). Journals that send papers out for review are called *refereed* periodicals. Some journals have higher standards than others and are more credible sources. Consult an experienced researcher to find out the most reliable and credible sources in your area of interest.

Books of Readings

Collected articles can be a mixed bag in terms of credibility. There can be classic papers that have been through the peer review process, have stood the test of time, and are printed in their entirety. Sometimes an original journal article is abridged or adapted for inclusion in a book of readings. There may also be some new chapters that have not passed through peer review.

Books on Behavioral Science Issues

A research library contains numerous books on behavioral science issues. Some are textbooks written primarily for specific courses. There are technical books and monographs directed to advanced specialists. There may also be some popular books written for a general audience. Some of these will be authored by well-known scholars who want to reach a general audience, or they may be written by professional writers specializing in behavioral science issues.

The important thing to remember about virtually all books is that they have not been peer reviewed. A publisher typically sends a book manuscript to several reviewers for their comments, but the major concern is likely to be the potential market (will it sell?), which is very different from the type of evaluation given to a journal article.

Each of the preceding outlets is intended to reach a specific audience. A researcher who studies animal behavior will publish articles in refereed technical journals to be read by colleagues. To reach a general audience, it is entirely proper and sensible for an animal behaviorist to write a nontechnical paper for *Natural History, Audubon,* or the *Smithsonian,* all of which are nonrefereed but still have high editorial standards. If the findings were particularly interesting, a commercial publisher might ask the author to prepare a popular book for a general audience.

Using the Library

Set aside specific time for your library work, and do it far in advance of the deadline for your paper. You might not be able to find everything you need on a single trip, and it will take a while to become familiar with the various information sources.

Keeping track—many researchers keep track of references on 3 × 5 index cards. For convenience, these are put in alphabetical order to be used for the reference section of the paper. If this is a one-time project or if you are prone to losing index cards, keep track of your references in a notebook. List them by author's last name, with first name and initials, additional authors, year of publication, title of book or periodical (with volume number and pages), and, for books, the publisher (Figure 3–1). It is very frustrating to finish the body of your report and find you are missing the reference information for an important study. Keeping thorough notes from the beginning will save a lot of time during the write-up. As you read materials, you can cross-reference your notes to your reference list by author's name and year of publication (e.g., Denney, 1985). If you are working on a computer you can make your own "card" file using your word processing or data base management program.

The most important and useful information resource in the library is not found on the shelves, but is sitting at a desk waiting to assist you. The reference librarian is a trained professional whose job is to help people gain access to information. Don't just ask and run; explain your needs in detail. There are many ways in which a librarian can help.

Consider the depth to which you want to explore the topic. For a broad overview, and to get a feel for previous research on the topic, the index of a general textbook in the field is a good place to start. There are also specialized encyclopedias in the library that will present concise descriptions and journal

Goldstein, A. B. & Pentz, M. A. (1984). Psychological skill training and the aggressive adolescent. *School Psychology Review*, *13*, 311–323

LB 1051
S 373

book

Csikszentmihalyi, M. & Larson, R. (1984). *Being adolescent: Conflict and growth in the teenage years*. New York: Basic Books.

HQ 796
C 89
1984

chapter in book

White, J. W. (1983) Sex and gender issues in aggression research. In R. G. Geen & E. I. Donnerstein (Eds.) *Aggression: Theoretical and empirical reviews*. (Vol. 2, pp. 1–26). New York: Academic Press.

BF 525
A 3
A 525

FIG. 3-1. Example of handwritten references.

definitions of topics in a number of fields. For example, the *Encyclopedia of Psychology* has brief sections on marriage counseling, repression, social learning theory, introversion–extroversion, character development and so on. There are also encyclopedias of Education, Social Work, Women's Studies, and many other fields. Finding your topic in a textbook or encyclopedia will ensure that it is, in fact, what you think it is. There are college students who have enrolled in a course on the *Psychology of Women* thinking it would help them in finding dates. Every field has its own jargon (specialized language) and it often differs from common usage.

Your report will probably require more information than is available in general sources such as encyclopedias and textbooks (often referred to as *secondary sources*). You will probably need to review both periodicals (journals) and books. These require two different paths of investigation. In both instances, it is wise to begin with the most *recent* publications, and then to work back to earlier sources, as needed.

Finding Books on Your Topic

The library will have a card catalog or computerized catalog, also called an *on-line catalog,* with subject headings. Most libraries use the subject headings from the two-volume *Library of Congress Subject Heading Index.* These are very large red volumes that are generally conspicuously placed in the card catalog area or reference room. As an example of their use, assume you are interested in alcohol-related accidents. That heading is **not** one of the Library of Congress Subject Headings, and therefore you won't find it in the card catalog. However, under the Library of Congress listing for alcohol, you will see NT (for Narrower Terms) with the following listings:

Drinking and airplane accidents
Drinking and automobile accidents
Drug–alcohol interactions

These should be the subject headings in the card catalog that contain a catalog card for each book available on the topic.

There is a considerable amount of information on the individual catalog cards (Figure 3-2). The call number (usually located on the upper left corner)

HE
5620
D7
D4
1986

Denney, Ronald C.
Alcohol and accidents / Ronald C. Denney. -- Wilmslow, Cheshire : Sigma Press, 1986.
viii, 172 p. ; 21 cm.
Includes bibliographical references and index.
ISBN 1-85058-067-7

1. Drinking and traffic accidents.
2. Drunk driving. I. Title

FIG. 3-2. Example of an entry in the main card catalog.

is essential for locating the book. It is also a good idea to jot down the author's last name. If the book turns out to be relevant for your paper, you will need to obtain additional information for your reference list—that is, names and initials of authors, year published, title, where published, and publisher. If you are using a chapter in an edited book, note the chapter title and page numbers.

There are differences among libraries in the degree of access to the main book collection. Libraries with "open stacks" allow readers to browse through the collection and select what they need themselves. Such libraries often have tables in the stack area for people to sit down and read books and articles on the spot. Libraries with "closed stacks" prohibit many readers from entering the collection area. The books and journals must be requested at a central desk and then retrieved by the librarians.

If your library has an open stack policy, check the library shelves around the call numbers that were mentioned most often in the card catalog. Librarians tend to place books on a topic close together. You are likely to discover titles that were missed in the card catalog search.

In addition to the card catalog, many libraries will have an on-line retrieval system, that is, a computerized card catalog. Most of these systems use the Library of Congress subject headings. Your library will have instructions on how to use the system. Since it can be done on a screen at a terminal, it is faster than sorting through drawers and cards.

Finding Journal Articles on Your Topic

Librarians refer to journals as *periodicals.* There are a number of periodical indexes. These are the equivalent of the card catalog for books, and will probably be located in the library Reference Room. Usually there is a helpful reference librarian nearby.

To begin your search for journal articles, you should select some *key terms,* like the subject headings you use to find books. For example, for eating disorders, you would use *eating disorders* as an index term, and also try *anorexia, bulimia, overeating, binge eating,* and so on. Look for these terms in different indexes (data bases). Librarians refer to these as specialized tools—sets of bound volumes that help you find the research that has been done on your topic. Some of the indexes have an accompanying *Thesaurus,* a listing of subject headings. Others do not, in which case you try your key words.

Social Sciences Index

A good place to start looking up key words is the *Social Sciences Index.* This provides a broad coverage of topics in all the social sciences. Each volume covers a specified time period. This index does not have its own Thesaurus.

Instead, the headings are similar to the Library of Congress listings. Take the most recent volume, look up your key words, and see the articles listed. Citations are given in terms of author(s), date, title, periodical, volume number, and pages (Figure 3-3). Write down the necessary citation information for relevant articles.

The next step is to obtain the call number of the journals containing the articles you wish to read. The call numbers will be located in the card catalog (as are book call numbers), or your library may have an additional listing in the Reference Room. In some libraries the call numbers are on microfiche (rectangular cards inserted in a monitor and read on a screen).

Psychological Abstracts and Sociological Abstracts

These indexes are more specialized than the *Social Sciences Index.* Their advantage is that in addition to the citation, there is a brief summary of the journal article (the abstract). From the abstract, you can tell whether or not the article will be of use. Use the appropriate *Thesaurus* (one for *Psychological Abstracts* and another for *Sociological Abstracts*) to check your keywords. You may find other useful key terms.

ERIC

ERIC stands for Educational Resources Information Center, and is a data base for educational literature. ERIC provides citation information and abstracts.

Keyword "Drinking and traffic accidents"
The alphabetical subject listing in the *Social Sciences Index* for April 1988 to March 1989 shows the following entries:

Drinking and the aged *See* Alcohol and the aged
→ **Drinking and traffic accidents** *See* Drunk driving
Drinking and women *See* Alcohol and women

Looking under **Drunk Driving**:

Drunk driving
See also
Mothers Against Drunk Driving
Alcohol, trauma and traffic safety; research issues, needs and opportunities. M. W. B. Perrine. *Contemp Drug Probl* 15:83-93 Spr '88
Arrests of women for driving under the influence. E. R. Shore and others. bibl *J Stud Alcohol* 49:7-10 Ja '88

(Additional entries are provided)

FIG. 3-3. Example of a citation in the Social Sciences Index.

Index Medicus

Index Medicus provides references to articles from biomedical journals and some social science periodicals. There are no abstracts in the written volumes, but these can be found in the computerized data base, **Medline.**

Social Science Citation Index (SSCI)

This three-volume index covers social science journals. It is available in bound volumes and on computer, but does not provide abstracts. There is a Subject Index, a Source Index, and a Citation Index. The headings listed in the Subject Index are derived from key words taken from article titles, so try all of your key words. Under the subject headings, only the authors' names are listed. To get the full citation, you must turn to the Source Index. There, under the author's name, you will find his or her publications for the specified time period including a list of the authors cited in the article. For example, if you were interested in teenagers and shopping malls, the 1981–1985 SSCI lists under the heading of **mall—shopping** the name, Anthony, KH (it is also listed under shopping—mall). Turning to the Source Index, you would find the citation for Kathryn Anthony's article "The shopping mall—a teenage hangout" and her reference list, published in *Adolescence* in 1985.

The Citation Index lists *all* of the past published articles by an author that were cited by someone for the period covered. If no one referenced a paper during that time, then the citation to that paper would not appear under the author's name. To give another example, you might know that Pollitt, Saco-Pollitt, Leibel, and Viteri published an article on iron deficiency and behavioral development in children in 1986 in the *American Journal of Clinical Nutrition.* Because you are interested in this topic, you would like to learn about more recent work. Since you have the original citation you don't need the Source Index. Instead, go to the Citation Index and look up Pollitt for a more recent year—for example, 1988—and you will find the following:

Pollitt E			
86 Am J Clin N	Vol	Pg	Yr
Watts, DL J ORTHO MED	3	11	88

This shows that Watts in 1988 cited the Pollitt article. Note that the title of Watts' article is not given. For that, you need to go the Source Index for 1988, and there you will find under Watts DL the article title, "The nutritional relationships of iron." If you are interested in behavioral aspects of iron deficiency, you may decide that this article is not relevant and not bother to look it up.

In this way you can find other researchers who may have published on your topic. The abbreviated journal listings are defined in the beginning of the index. For example, PSYCHOS MED is *Psychosomatic Medicine,* J PERS SOC is *Journal of Personality and Social Psychology,* and so on.

Other Specialized Sources

Sage Race Relations Abstracts
Women's Studies Abstracts
Social Work Research and Abstracts
Child Development Abstracts
Criminology and Penology Abstracts
Exceptional Child Education Abstracts

Your library may have others not mentioned here.

Less Technical Sources

Sometimes the information you want will be in the mass media—newspapers and magazines. If so, use the *Reader's Guide to Periodic Literature* and the *Popular Periodical Index.*

Computer Retrieval Systems

Many of the preceding indexes are also available on-line; that is, by computer. A computer search of these indexes can speed up the process (Figure 3-4). If time and money are available, ask the reference librarian about a computer search. You often have to pay for this service. Be prepared to provide three or four key words describing your topic. On-line systems are available for *Psychological Abstracts, Sociological Abstracts, ERIC,* and *Index Medicus* (on-line system is called **Medline**). There are data bases for the criminology literature.

It is likely that some data bases will be available directly to you at your library, a situation requiring time rather than money. Learning to use the on-line system requires an investment of a few hours. If time is short and you are not comfortable with data base systems, you are probably better off using the shelved volumes. Having done it "by hand" will facilitate subsequent learning of the on-line system. The printed indexes usually provide a more complete range of publication years, going back farther in time. Bound volumes of *Psychological Abstracts* extend to the 1920s.

Title	TI: Deterrence of Drinking-Driving: The Effect of Changes in the Kansas Driving under the Influence Law
Author	AU: Shore, -Elsie-R.; Maguin, -Eugene
Institution	IN: Dept Psychology Wichita State U, KS 67208
Journal	JN: Evaluation-and-Program-Planning; 1988, 11, 3, 245-254

The additional information on citation is not necessary for your search. An abstract, termed AB, is also provided.

FIG. 3-4. Example of an on-line citation from Sociological Abstracts.

Other Strategies

In addition to searching by topic, you can search by name. If you have identified key individuals who have done work on the topic, use the author index in the various sources to find listings of their work.

For the most recent publications (within the past 6 months), make a note of journals that were cited most often in your search and in the books you found on your topic. Then visit the Periodical Room. It houses current issues of journals, magazines, and newspapers. Find the current issues of the journals that focus on your topic and skim their tables of contents for current articles. Earlier volumes will probably be on shelves in another part of the library.

How Far to Search?

Begin with the most recent work. If the topic is limited to a specific discipline such as psychology, then using the card catalog for books and *Psychological Abstracts* for journals will probably be sufficient. If you are researching an interdisciplinary subject—for example, recreation and leisure—you may need to use more than one data base.

How far back in time you should go will depend on the amount of recent material you find. If there are many recent articles, these will probably contain reviews or summaries of earlier work, and save you the trouble of having to read through all the earlier papers. If the earlier work is especially close to your interest, you will probably want to read the original paper rather than rely on someone else's interpretation. If there is very little current research, then you will have to go further back in time to find out what is known about the topic. The key is to find the most relevant articles. How many will depend on the amount and quality of published research.

Sometimes you will be fortunate enough to be doing research on a topic for which a literature review is available. There may be a *bibliography,* a compilation of titles of earlier work on a topic. A useful library reference tool is the *Bibliographic Index,* a topical index published yearly. Some bibliographies are published separately, and others may be buried within books or articles. The *Bibliographic Index* provides the exact location, even to the page numbers, of the bibliography. Some government agencies periodically publish bibliographies of research studies in particular fields. There also may be review articles which include critical evaluations. A few journals in the behavioral sciences, such as *Psychological Bulletin,* specialize in literature reviews. On-line computer systems such as *Psychological Abstracts* and *Medline* allow the user to search specifically for reviews. If you are fortunate enough to be working in an area where a previous review exists, you will still need to update the material. Because of the publication lag, a review article published in 1990 is probably current only through 1988. There may be additional sources or references not included in the review.

When there are no published literature reviews or bibliographies, you can obtain information by networking. Look at the introductory sections of published papers on the topic and see which studies are cited. A single source can supply certain bits of information and point you toward other sources. When the names and titles that you encounter begin to look familiar, then you have come close to a good overview of the area. Your information on a topic will never be complete. Although it is desirable to learn as much as you can before the study, there will always be new bits of information coming in while the project is underway.

Interlibrary Loan

Some books or key articles turned up in your literature search may not be available at your library. The book may be missing, the journal may have been sent out for binding, or the library does not subscribe to the particular periodical. When an unavailable article or book seems particularly important, consider using *interlibrary loan.* The librarian who handles these requests has computerized lists of materials at other libraries. Typically, you fill out a request card for each book or periodical needed. When it arrives, the interlibrary desk sends a postcard telling you to pick it up and how long you can have it. For brief articles, the situation is somewhat easier, since the library that subscribes to the periodical can make a photocopy and send it for you to keep. The chief problem with interlibrary loan is the time delay. Several weeks can elapse between the initial request and the arrival of the book or journal.

Government Publications

Government documents such as census reports and agency publications may not be listed in the card catalog. They will be kept in a special documents section with its own librarian, who can locate the relevant reports. Some that are not in the library can be ordered from the Government Printing Office for a nominal charge. Documents can also be borrowed from other libraries or directly from a government agency.

What to Look for in a Research Article

Your first screening is the title of the article and key words. If you are studying the effects of television violence upon aggression in children, you would look for "television," "aggression," "violence," and "children's aggression." At least one of these terms should be present if the article addresses that topic. Read the abstract. If it looks promising, then locate the article. If it is

BOX 3-1. Steps in a Library Search

1. Look up topic in textbooks and encyclopedias.
2. Check card catalog under subject heading. Write down call numbers of relevant books.
3. Check with reference librarian.
4. Consult periodical indexes by topic (after checking Thesaurus), and by authors (if known) who do research on the topic.
5. Reference cards or lists should include the following information:
   A. For **books**
      1) Last name and initials for all authors
      2) Year of publication
      3) Title, including edition or volume number, if appropriate
      4) Place published, city and state/province
      5) Publisher
   B. For **book chapters**
      1) Last name and initials for all authors
      2) Chapter title
      3) Page numbers
      4) All book information as in A1–5
   C. For **journal articles**
      1) Last name and initials for all authors
      2) Year published
      3) Title
      4) Periodical
      5) Volume number
      6) Page numbers
6. Consider the value of a computer search.
7. Scan table of contents in current issues of journals on your topic.
8. Consider other possible information sources (e.g., government documents, inter-library loans.

pertinent, read it through, section by section, keeping in mind the following advice.

Some studies are of better quality than others, and should be given more weight in your review. You need to know how to distinguish between good and less-good studies. You also need to know what information is more important and what is of lesser importance in order to take notes. To some extent, this will come with experience. To get a general idea of what to look for, skip ahead to Chapter 20 and look at Box 20-3. Key points of information that should be included in a journal article are listed. These are also the points that you may wish to cover in your brief notes for the introduction section of your paper (i.e, what the researchers were looking for, what they did, and what they found).

If your topic is a broad one, such as sex differences, you will need to make it more narrow and be selective in your review. More manageable topics would be gender stereotypes in the media, sex differences in aggressive behavior, or children's perceptions of sex roles.

In reading an article do not overlook the Method and Results sections. These are often the distinguishing features between good and mediocre studies. The method determines the reliability and validity of the findings. The results section presents the findings without subjective interpretation. Some journal formats combine the Results and Discussion, requiring a more critical reading to separate findings from interpretations.

As the prose is more tightly packed with information, reading a technical article is slower than reading a newspaper or novel. There are dictionaries of behavioral and social science terms and notations that are available in most reference libraries. The glossary of this book and of statistical textbooks will contain definitions of the major terms connected with methodology and statistics.

Direct Consultation

Find out the names of local individuals and agencies knowledgeable about the topic. They can supply information specifically about the situation in the community that may be included in a report. They can also be asked for the names of knowledgeable individuals.

It is gratifying to find how helpful outside people can be if they are approached in the right way. A further advantage of local consultation is the possibility of making valuable contacts. During conversations with local officials, students may learn about new programs and job opportunities.

Distant officials and authorities are more difficult to use as information sources. If you write to the author of a book on a topic, make your requests specific. Some famous people are extraordinarily helpful with student inquiries. If you have a specific question that you think some distant authority can answer, don't hesitate to write a personal letter. Enclosing a stamped self-addressed envelope will increase the likelihood of a reply.

Summary

Reviewing the literature means finding out what previous research has been done on a topic. Primary sources such as journal articles and books by scholars provide the most credible information on a research topic. These can be located in reference libraries using the *Library of Congress Subject Heading Index* as a guide to topics covered by books. Periodical indexes such as the *Social Sciences Index* and *Psychological Abstracts* list citations for journal articles. Do not hesitate to enlist the help of the research librarian. Begin the search with the most recent publications, and either narrow or broaden your focus, depending on the information available. There will also be situations in which direct consultation with known authorities will be helpful. Allow sufficient time for finding the necessary materials.

Further Reading

Douglas, Nancy E., & Baum, N. (1984). *Library research guide to psychology: Illustrated search strategy and sources.* Ann Arbor, MI: Perian Press.

Greenwood, Larry (1980). *How to search for information: A beginner's guide to the literature of psychology.* Lexington, KY: Willowood Press.

Jolley, J. M., Murray, J. D., & Keller, P. A. (1984). *How to write psychology papers: A student's survival guide for psychology and related fields.* Sarasota, FL: Professional Resource Exchange.

Kennedy, James R., Jr. (1979). *Library research guide to education: Illustrated search strategy and sources.* Ann Arbor, MI: Perian Press.

McInnis, R. (1982). *Research guide to psychology.* Westport, CT: Greenwood Press.

McMillan, P., & Kennedy, J. R., Jr. (1981). *Library research guide to sociology: Illustrated search strategy and sources.* Ann Arbor, MI: Perian Press.

Reed, Jeffrey G. (1982). *Library use: A handbook for psychology.* Washington, DC: American Psychological Association.

Rosnow, R. L., & Rosnow, M. (1986). *Writing papers in psychology; A student guide.* Belmont, CA: Wadsworth.

II
Observational Methods

4 Observation

In one of the earliest social psychological experiments, Triplett (1897) found that people wind fishing reels faster when they are with other people than when they are alone. This *social facilitation effect* has been found in other laboratory studies with people and with nonhuman species. In some studies, after an animal was fed all it wanted to eat, another animal was brought into the room to be fed. At this point the first animal would resume eating.

An important issue from the standpoint of external validity is whether these effects are found among people in their daily activities—for example, do people eat and drink more when they are in groups than by themselves? This question is well suited to the unobtrusive observation of people in public settings. Earlier researchers enthusiastically used this approach in bars and cocktail lounges. They found that people in groups consumed more alcohol than people alone. However, groups did not drink faster, as had been popularly supposed (slow individuals rushing to keep up with fast drinkers), but rather because groups stayed longer than did lone individuals, and length of stay was strongly related to alcohol consumption.

The authors of your textbook were interested in seeing if facilitation effects occurred in coffeehouses. We recruited three student observers, trained them in observational methods, and assigned them to different coffeehouses. The first coffeehouse was used primarily by business people and office workers, a second was in a shopping center, and a third was in the student union. Each observer entered the coffeehouse, bought a cup of coffee, and sat down at a table with a good view of the surrounding area. The student tried to blend in, appearing to be a student reading a textbook and taking notes. Each session took 2 hours or until the observer filled the quota of observing three lone individuals and three people in groups.

In all three coffeehouses, groups stayed longer than lone individuals, and people who were joined stayed longest. Facilitation effects were much more apparent in duration (length of stay) than in beverage consumption, since most patrons only had a single cup of coffee. The instructors who supervised the project published their results in a technical journal (Sommer and Sommer, 1989), bringing recognition to the student observers whose contributions were acknowledged in the article.

For those who enjoy people-watching, observation is the ideal research method. It is useful not only as a method in its own right but as an accompaniment to other procedures. Before beginning an interview study, it will be necessary to observe the situation first. You will want to know where to find people, how long they are going to be available, and possible distractions. Before beginning an experiment, you will want to know about the behavior in its natural state. Otherwise you run the risk of creating conditions in the laboratory that do not exist in the real world.

An advantage of observation is that it does not require conversation. You can observe pedestrians on a sidewalk 20 stories below or people who do not speak your language. Observation is the ideal method for studying commonplace nonverbal behaviors, such as gestures, postures, or seating arrangements, in which people may not be consciously aware of how they are acting.

Observation is economical in terms of money and equipment but expensive in terms of time. One invests long hours of waiting. This is well known to animal researchers, who spend weeks scouting the terrain and establishing observation posts before catching a glimpse of a rare bird or mammal. More than any other method, observation requires patience and luck. There is no certainty about who will appear and what will happen. It may rain or snow while you are there, be too hot or too cold, the setting too crowded or too empty. Also, without expensive equipment and apparatus, the observer's skills are severely tested. On the other hand, it is intensely satisfying to notice things that other people overlook. Generalization to the real world is easier when you have studied natural behavior. There are three types of observational procedures—casual observation, systematic observation, and participant observation.

Casual Observation

Casual observation is done without prearranged categories or a scoring system. It refers to eyeball inspection of what is happening. It is most useful at an early stage of research or as an accompaniment to some other procedure. For example, before one approaches hospital patients for interviews, it is desirable to spend some time watching behavior on the ward. This will yield information that is indispensable for developing good questions. Casual observation is not a substitute for more systematic and detailed study, but occasionally it is the only method possible. For example, a city planner visiting a housing project in India may have only a limited time to look around. This may mean a single tour in the company of a guide without the opportunity to talk to local residents. Keen powers of observation are required under these circumstances. Unless one knows what to look for, one might as well buy picture postcards or a tour book and save the time and trouble of a personal visit.

The vivid impressions of a first visit are worth recording. After several sessions, it is easy to become accustomed and desensitized to what is happening. First impressions are most useful when they are written down immediately. There can be no substitute for field notes kept on a day-by-day basis. These notes need not be typed or written in perfect grammar. A final report

Casual observation is an ideal technique for people watchers. Observing the people in a zoo can be as interesting and rewarding as watching the animals.

is best written after a long period of reflection, data gathering, and several drafts, but first impressions are most valuable when written while they are still fresh.

The notes in Box 4-1 were written by an observer following an hourlong session in a Canadian pub. The observer scribbled notes at the bottom of the observation sheet and typed them when he arrived home. In his notes, lone drinkers were referred to as Sols (solitaries). The notes were left in rough form because they were not primary data.

Systematic Observation

There are four major differences between observational research and just looking (Bickman, 1976).

1. It serves a specified research purpose.
2. It is planned systematically.
3. It is recorded systematically and related to more general propositions rather than simply being presented as a set of interesting curiosities.
4. It is subjected to checks and controls on validity and reliability.

Systematic observation employs a scoring system and prearranged categories that are applied consistently. This usually requires an observation checklist, on which information is recorded under the proper headings. Categories on the checklist should include those items of behavior that occur naturally in the situation and can be observed and recorded. Not everything that takes place is open to view. Casual observation is helpful for developing the categories to be used in systematic observation. In the study of drinking patterns in 15 bars in a New England city, Kessler and Gomberg (1974) found that they could record the following items on napkins or paper sheets without appearing conspicuous.

Number of drinks	Alone or in a group
Type of drink	Type of clothing
Number of sips per drink	Height estimate
Time to consume	Weight estimate
Total time in bar	Age estimate

The items were limited to what could be seen directly. What the patrons said to one another, their marital status, and their political attitudes lay beyond the range of an observational study. In establishing the reliability of the observational categories, two researchers went to a selected bar, sat either at a table or at the bar with instructions to independently observe the next patron entering the premises who sat in clear view of both observers. The patron was then observed throughout his or her stay in the bar. Agreement was high between the two observers using this procedure (almost 100 percent) except in the case of the age estimates.

BOX 4-1. A Visit to the King Edward Hotel

The King Edward is slightly out of the way and this may have had some effect on the patronage, because on this particular Saturday evening there were very few people around. The major cause of the low amount of people was that this was the Saturday of the only holiday weekend in August and many people were out of town. One could also say that the King Edward depends on its daily rather than nightly group of patrons for the major portion of its customers. In the afternoon there were men in jackets and such; many white-and blue-collar workers and laborers but not the large numbers of transients such as would be found in bars closer to the main drag.

The TV was also on in the bar that evening. Every solitary in the bar was watching it, which undoubtedly has some bearing on the figures. Neither of the groups was watching it. No one was reading. There are two TV sets in the place, one facing toward the upper level of this two-level bar; however, there were no people sitting on the upper level. (There was no restriction against sitting on the upper level, it just wasn't done.) The fact that the TV was on may have had some effect on the seating pattern, more people sitting in the center where they could see the TV without distortion. The rear wall would be a considerable distance from the TV screen, and as the screen was not of a larger than normal size, the viewers in the center were quite close to the set that was located above the serving counter. Two solitaries who entered after a new program had gotten underway after 10 sat to the far side where they could neither see the screen nor be easily seen by me.

Sol A, a watcher, was an elderly man with his back to me, was watching intently, and sipping quite regularly, not looking to see the glass before he lifted it. He was, I would say, drinking unconsciously. I also noted he waited until the program changed before he got up to go to the lavatory.

Sol B was an elderly man in work clothes of a clean nature. He could perhaps be a night watchman, or something of that line. I noted he looked at the clock quite frequently in the 15 minutes or so he was there (I was there for about 5 minutes or so before beginning notations) and left at 9:35 before the program was over.

Group A was an almost nongroup group. There was very little conversation, and very little interest in anything beyond the table. "A" was dressed in sports clothes: checked cotton shirt, low-cut shoes. They weren't watching TV and left in the middle of the program.

B was in the large group. Conversation at this table ranged through the realm of sports with reference to the case before the courts of whether bridge is a game of chance or a game of skill, and whether Edmonton would have a good team in this year's football. The table was mixed—middle-aged and old, sports clothes and shabby, large glass and small. Five of them referred to family and children at one time or another, so this might be a boys'-night-out group of some sort. The extremely interesting aspect of this is that when three of the members of the group left there were only large-glass drinkers left. This could be coincidence or it might be significant. The three who left were all small-glass drinkers.

The observer must choose a location from which behavior can be seen and recorded. This will be easier in some settings than in others. In the coffeehouse, it was possible for the observer to station herself at a table to record activities. A city planning student studying behavior in a city park found a good view from a tall building overlooking the park (Ciolek, 1978). The choice of an observational post depends completely on local circumstances. There is no way to know in advance what the best vantage point will be. A videocamera can be useful in observational research. Interactions can be recorded at one time and transcribed later (see Chapter 14).

How to Be a Good Observer

The first principle of good observation is to heed the Greek maxim "Know thyself." Careful attention to your own responses provides valuable insights into what is happening in the situation and in you. Make it a practice to acknowledge and *name* your feelings: "I am beginning to become tense . . . I feel comfortable . . . something seems odd here." Try practice observations with a friend and share your feelings aloud. See if you can identify the internal cues for these feelings. Notice whether you hold your breath or stand differently when you are tense. Practice plus feedback will increase your sensitivity to such feelings.

Pay attention to nonverbal cues in the environment. These include people's posture, gestures, and privacy-seeking behaviors, such as turning away and gaze avoidance. When an anthropologist was observing a mental hospital ward, he found that patients acted disturbed when he sat out in the dayroom, and the nurses were bothered when he was inside the nurses' station. He gradually found himself forced back into a small area outside the nurses' station. Monitoring other people's reactions to his presence taught him which places were open and closed to him.

Systematic observation requires the researcher to take notes in an unobtrusive manner. The closer the recording procedure resembles normal activity in the setting, the better. One observer in a restaurant recorded observations on napkins; another in a classroom recorded participation in a notebook. Certain locations are more suitable than others for recording what is happening. Watching bicyclists at an intersection proved difficult when the observers stood on the street corner; cyclists became curious and altered their behavior. It was more effective to watch from inside a parked car. The researcher studying a playground might bring along a young child as a cover, to make the oberserver's presence more comprehensible. "Fitting in" is an art that develops through practice. The objective of the naturalist is to disturb nature as little as possible while studying it. Most of the time, it pays to have an observer who can blend into the setting. In laboratory research, it is often assumed that the physical appearance of the experimenter does not matter. Whether this assumption is true is an interesting question, but no one ever assumes it in a natural setting. Researchers who get their first introduction to observational procedures in a setting in which they do not fit will probably soon become discouraged. Elliott Liebow, who spent a year among streetcorner men in a black neighborhood in Washington, D.C., realized in the end that he would always be an outsider:

> This brute fact of color, as they understood it in their experience, and I understood it in mine, irrevocably and absolutely relegated me to the status of outsider. . . . I used to play with the idea that maybe I wasn't as much of an outsider as I thought. Other events, and later readings of the field materials, have disabused me of this particular touch of vanity. (Liebow, 1966, p. 248)

George Schaller (1964), who studied gorillas in their natural habitat, describes the gradual process of becoming a good observer in the jungle:

> In civilization, one loses the aptitude for stillness, the habit of moving gently. It takes time to cease to be an outsider, an intruder, and be accepted once again by the creatures of the forest. The return to the wilderness is a gradual process, unconscious for the most part. Once the senses have been relieved of the incessant noise and other irrelevant stimuli that are a part of our civilization . . . the sights, sounds, and smells of the environment become meaningful again. Slowly the courage and confidence of man, previously nurtured by his belief in the safety of his civilized surroundings, slips away. Finally he stands there, a rather weak and humble creature who has come not to disturb and subdue but to nod to the forest in fellowship and to claim kinship to the gorilla and the Sunbird. (1964, p. 107)

In natural observation the researcher refrains from interfering in the ongoing activities. When a researcher decides to intervene—for example, by changing the furniture in a room or placing a red ball in a forest clearing—the procedure is known as a field experiment. The chief exception to the rule of noninterference occurs in participant observation, where the observer is part of the event being studied.

Those who observe behavior in natural settings can expect to endure physical discomfort and occasionally danger. It is not uncommon for zoologists studying animal behavior to stake out an area and spend weeks in fruitless waiting in a stuffy, hot, insect-ridden blind. Watching children play in the streets of Philadelphia during July and August may be no less uncomfortable and occasionally more dangerous. Box 4-2 lists some of the potential problems in systematic observation.

Reliability in Systematic Observation

In systematic observation it is desirable to employ at least two independent observers at an early stage of study. Independent observers are two or more individuals who take notes separately and compare them afterward. Such comparisons may reveal ambiguities and overlaps in scoring categories. These comparisons should be made before the main body of observations is collected, not after. After you have finished making all your observations, if you learn then that some of the categories are not reliable, the value of the

BOX 4-2. Pitfalls to be Avoided in Systematic Observation

1. Reactive effects from being observed; a guinea pig effect in which awareness of being watched changes behavior.
   a. People becoming self-conscious and not behaving as they normally would.
   b. People attempting to accommodate the observer, doing what they believe the observer wants them to do.
   c. Influence of the observer's specific appearance or manner on people's actions.
   c. Changes in accommodation to the observer during the course of the study.
2. Investigator error
   a. Unclear and unreliable observational categories.
   b. Bias on the part of the observer.
   c. Changes in the observational procedures in the middle of the study.
3. Sampling error
   a. People being observed are not representative of the groups to which the results will be generalized.
   b. Inadequate time periods selected for observation.
   c. Sources of bias due to weather, day, location, etc.

data will be diminished. When such problems can be identified beforehand, the scoring system may be improved to eliminate them, and then reliabilty checked a second time so that the observations can begin with confidence that the scoring system is reliable. Box 4-3 lists the steps for doing systematic observational research.

BOX 4-3. Steps for Systematic Observation Research

1. Specify the question(s) of interest (reason for doing the study).
2. Do casual observation, distinguishing between observation (the actual behaviors seen) and inference (interpretation, what you think it means).
3. Are the observational categories clearly described?
4. Design the measurement instruments (i.e., checklists, categories, coding systems, etc.).
5. Is the study designed so that it will be *valid,* (i.e., does it measure what it is supposed to measure and have some generalizability)?
6. Train observers in the use of the instruments.
7. *Do a pilot test.*
   a. Test the actual observation procedure.
   b. Check reliability of the two categories using at least two independent observers.
8. Revise procedure and instruments in light of the pilot test results. If substantial changes are made, run another pilot test.
9. Collect data.
10. Compile, analyze, and interpret results.

Qualitative Approaches

The two observational methods described so far, casual and systematic observation, involve an outside observer coming in to watch and record what is happening, often for brief periods according to a time sampling system. Qualitative approaches put more emphasis on detailed personal description and less on categories and quantification. These approaches are time-consuming and place heavy responsibility on the observer to maintain detailed records. Two types of qualitative observation are *participant observation* and *ethnography.*

Participant Observation

In *participant observation* the observer becomes part of the events being studied. The emotional learning on the part of the researcher can be as important as the documentation of external events. One participant observer described what he had learned over a 6-month period, "It wasn't a question of discovering new facts, since most of what I had found was already known, but of discovering what it meant to *feel* the facts."

Criminologist George Kirkham took a year's leave from his university position to go through training in a police academy and work as a patrolman. Many of his fellow officers knew of his research interests, but the people on his beat looked on him as an ordinary policeman. Kirkham writes with great sensitivity about the changes in his own attitudes that occurred when he worked in a high-crime neighborhood. His records included events that occurred on his beat, his own attitudes and those of his fellow officers, and the responses of his family and friends to his new role.

> According to the accounts of my family, colleagues, and friends, I began to increasingly display attitudinal and behavioral elements that were entirely foreign to my previous personality—punitiveness, pervasive cynicism and mistrust of others, chronic irritability and free-floating hostility, racism, and a diffuse personal anxiety over the menace of crime and criminals that seemed at times to border on the obsessive (Kirkham, 1975, p. 19).

In his role of policeman, he became a vociferous advocate of capital punishment even though, as a criminologist, he acknowledged its ineffectiveness as a deterrent. He also resented the courts' "coddling" of criminals. His inability to reconcile his conflicting perspectives began to seem like hypocrisy. He began to have doubts about who was the real George Kirkham. The answer, he acknowledged, lay in the situations in which he found himself. It was much easier to be detached and objective around a seminar table with a group of students than as a policeman in direct contact with violence.

Sharing the risks with his fellow officers gave Kirkham insights that he felt could not have been obtained through interviews or questionnaires. The hyperaggressiveness and braggadocio of his fellow officers began to make sense as a collective defense mechanism. Any show of weakness on the part of one patrolman threatened the entire force. Kirkham left his beat after a year's service and returned to a placid existence as a university professor. Unanswered questions impelled him to embark on an additional 6-month stint as a policeman in a small community with very little crime. The tension and hositility of the previous job were absent. He could come home and take off his badge and uniform with a "new and far deeper feeling of service and satisfaction" (Kirkham, 1975, p. 19).

A participant-observer has a defined and active role in what is happening, as distinct from being a spectator, bystander, or customer. This is not to imply that customers cannot make observations. The coffeehouse study described earlier was done by observers posing as customers. This seems more properly classified as systematic observation rather than participant observation. Had the observer been a waiter or bartender who took the job specifically to record drinking patterns, then it would have been participant observation.

Researchers sometimes find themselves in a participant observation role through circumstance—for example, after an automobile accident or physical disability. Research through participant observation is a means of understanding the experience and also a way for them to use their professional training while they are immobilized.

Ethnography

The description and study of specific peoples and places is known as *ethnography.* At one time, ethnographers were concerned primarily with primitive societies. More recently they have turned their attention to the study of contemporary peoples in special settings, such as courtrooms, banks, and shopping malls. The observer looks, listens, asks questions, and records what is seen and heard. From all this emerges a picture of what is happening. This is not the picture of the artist who sketches a scene from a single perspective. It is more the approach of the detective who examines and puts together all sorts of evidence, including smudged fingerprints, bloodstains, torn clothing, and eyewitness accounts (Sanders, 1973). Although the terms *ethnography* and *participant observation* are often used interchangeably, ethnography, like other related terms used in the social sciences, such as qualitative sociology and field work, is more of an approach than a research method, because it combines several research techniques, including interviews, observations, and physical trace measures. The approach is particularly useful in the early stages of research where little is known about a phenomenon, in situations where other methods are not feasible, and as part of a large multimethod strategy.

In both participant observation and ethnography, impressions are written down in rough form and may be typed later. The final form should be legible, with places, dates, and names (real or in code) indicated, and a clear distinction between events witnessed and the observer's impressions or feelings. It is quite useful in field notes to describe one's own interpretations, but these should be placed in brackets or otherwise distinguished from actual events. These records are later summarized and discussed with colleagues or a research supervisor. Collecting field notes is often done by a single researcher, but the subsequent validation of observations and interpretations is a shared enterprise.

Flexibility is a required attribute for a participant observer or ethnographer. Methods will evolve during the course of the research, and procedures will be added, modified, or dropped. Unlike a laboratory study, new sources of information will appear during the course of the investigation, requiring new approaches or contacts. During field work, it will be impossible to keep track of everything happening at one time. Some selection among different variables or processes will be necessary, along with constant monitoring of how they interact and measure in importance.

Kirk and Miller (1986) identify invention, discovery, interpretation, and explanation as the four phases of field work. *Invention* refers to preparation for the study and the development of the research plan. This includes reviewing previous work in the field and training in appropriate methods. *Discovery* refers to data collection, the production of information through a variety of methods. *Interpretation* is ongoing as the researcher reflects on what is happening and discusses it with colleagues, but it is also a separate phase following data collection in which there is formal analysis of what has been learned. *Explanation* refers to the packaging of findings for an outside audience. Typically this involves writing a report or article based on the study.

Reliability in Qualitative Research

The field notes of a single participant observer or ethnographer necessarily lack reliability. An observer may have blind spots for certain things that occur and exaggerate the importance of other events. Sometimes a participant observer gets so deeply into a role that perspective is lost, a process known as going native. The observer's presence may affect the behavior of the people in the setting. These effects are difficult to specify in advance and interpret. There are also problems in generalizability with this method—how much can be learned from the study of a single health clinic or police department?

Although it is difficult to eliminate these problems completely, it is possible to minimize their occurrence and effect on the research, and to specify when they have occurred. Most of the remedies involve *triangulation* or the use of more than one method, observer, and site to provide additional checks on a single observer's account (Hunt, 1985). Triangulation allows the researcher to pinpoint aspects of a phenomenon more accurately by approaching it from different vantage points using different methods (Brewer and Hunter, 1989). Successful triangulation requires careful analysis of the type of information provided by each method, including the strengths and weaknesses.

The likelihood of a researcher going native can be reduced by having a team of researchers in the field who interact with one another and thereby prevent total immersion in the local culture. Frequent contact between the researcher and the supervisor can also provide opportunities for reflection on what is taking place. Observer effects can be dealt with through a multimethod approach that uses independent sources of information, such as public or private records, previous research studies, and so on. When an observer's account contrasts markedly with what is known about the setting and the occupants, as when a participant observer finds no illegal gambling at a social club that has been raided by police on previous occasions and numerous convictions obtained on gambling charges, the researcher will have to seriously reflect on his or her impact upon the setting. Reliability can be improved by the use of more than one observer. This does not have to be done routinely. Only a few visits by a second observer may be sufficient to determine that the first observer is picking up the important things that are happening and reporting them correctly. Generalizability can be improved by increasing the number of settings observed. Instead of focusing all of the attention on a single setting, several representative locations can be observed. If the same behaviors are seen in all the settings, then the observer can have more confidence in the findings. Again this does not require a detailed observational study in all the settings. The additional sites can be observed briefly if the observer is not interested in everything that occurs, but only in specific items.

Stresses on the Observer

Of the techniques described in this book, participant observation is probably the most stressful for the investigator. The decision to use this method should not be made lightly. Undertaking field work among psychiatric patients living in the community, Estroff (1981) felt alone and isolated. There was no one with whom she could share the intimate facts of the daily observations and her own personal reactions, and no one to whom she could express her occasional feelings of frustration, depression, or even elation about clients, staff, their interaction, or her own reactions to a roller coaster set of experiences.

Some uses of participant observation, although not all, have been severely criticized both inside and outside the research community. Ethical problems can be minimized when all of the following conditions are met: (1) the researcher is qualified for the role being occupied, (2) the observer does not drain resources and services intended for others, and (3) the observer informs colleagues, co-workers, and clients of the dual role to the extent that this is practicable. As a participant observer, one is privy to backstage behaviors. It is virtually *inevitable* that one will see illicit, unethical, and illegal acts. A participant observer working in a restaurant, for example, is likely to notice unsanitary conditions. One could put this more strongly and say that the observer *will see,* over a period of time, unsanitary conditions. There is probably no restaurant in the world where this would not occur at least occasionally. How the observer deals with the situation is likely to be a source of severe personal stress. Should he or she "rat" on the other employees or keep quiet and risk exposing the public to a health risk? Ethical dilemmas are inevitable in participant observation.

There is also legal ambiguity about the status of field notes written by researchers. The courts have not determined that such field notes are protected against subpoena. They might become part of the public record in court proceedings.

Limitations

Observation deals with behavior, not with attitudes or beliefs. Attitudes can be deduced from behavior only with caution. If you want to find out what people do, you can observe them. If you want to find out what they think, you should ask them directly. There are exceptions to both of these rules, but observation is generally a good method for studying natural behavior, while interviews and questionnaires are more appropriate for opinions and beliefs.

Reliability is always a problem in observation. In casual and systematic observation, the use of two independent observers is recommended during the early stages of the study. No matter how simple and straightforward the behavior being studied, it is still wise to check on reliability. If two independent observers cannot agree on what they see, then the conclusions of the study are in doubt.

In qualitative observation, solutions to reliability problems involve triangulation, or the use of more than one method, observer, and site to provide additional checks on a single observer's account. The immersion of the observer into the situation for extended periods can be stressful. There are also potential ethical problems when the researcher becomes part of ongoing events.

Summary

Observation is useful in behavioral research as a method in its own right and as an accompaniment to other procedures. It is economical in terms of money and equipment but expensive in terms of time.

Casual observation does not use prearranged categories or a scoring system. It is most useful at the beginning stages of research. Systematic observation employs detailed categories and a scoring system. Possible sources of error in systematic observation are reactive effects from being observed (the guinea pig effect), investigator error, and biased sampling.

Two qualitative approaches are participant observation and ethnography. In participant observation the observer becomes part of the events being studied. Ethnography is the study of specific peoples and places. Both techniques place considerable demands upon the observer. There may be problems of reliability and generalizability. Limitations: Since observation deals with behavior, it is difficult to deduce beliefs, attitudes, or opinions. In many settings reliability is difficult to ascertain.

References

Bickman, L. (1976). Data collection I; observational methods. In C. Selltiz, L. S. Wrightsman, & S. W. Cook, *Research methods in social relations* (3rd ed.). New York: Holt, Rinehart and Winston.

Brewer, J. & Hunter, A (1989). *Multimethod research.* Newbury Park, CA: Sage.

Ciolek, T. M. (1978). Spatial behavior in pedestrian areas. *Ekistics, 268,* 120–122.

Estroff, S. E. (1981). *Making it crazy.* Berkeley, CA: University of California Press.

Hunt, M. (1985) *Profiles of social research.* New York: Russell Sage Foundation.

Kessler, M. & Gomberg, C. (1974). Observations of barroom drinking: methodology and preliminary results. *Quarterly Journal of Studies on Alcohol, 35,* 1392–1396.

Kirk, J., & Miller, M. L. (1986). *Reliability and validity in qualitative research.* Beverly Hills, CA: Sage.

Kirkham, G. L. (1975, May). Doc cop. *Human Behavior,* pg. 19.

Liebow, E. (1966). *Tally's corner.* Boston: Little, Brown.

Sanders, W. B. (1973). *The sociologist as detective.* New York: Praeger.

Schaller, G. B. (1964). *The year of the gorilla.* Chicago: University of Chicago Press.

Sommer, R., & Sommer, B. (1989). Social facilitation effects in coffeehouses. *Environment and Behavior, 21,* 651–666.

Triplett, N. (1897). The dynamogenic factors in pacemaking and competition. *American Journal of Psychology, 9,* 507–533.

Further Reading

Fetterman, D. M. (1989). *Ethnography step by step.* Beverly Hills, CA: Sage.

Hartmann, D. P. (Ed.). (1982). *Using observers to study behavior.* San Francisco: Jossey-Bass.

Jorgenson, D. L. (1989). *Participant observation.* Beverly Hills, CA: Sage.

Lofland, J. & Lofland, L. (1984). *Analyzing social settings: A guide to qualitative observation and analysis* (3rd ed.). Belmont, CA: Wadsworth.

Mellen, J., Hage, S., Pfeifer, S., & Carlson, D. (1983). *Research methods for studying animal behavior in a zoo setting* [Videotape and text]. Apple Valley, MN 55124: Minnesota Zoological Garden.

Punch, M. (1986). *The politics and ethics of fieldwork.* Beverly Hills, CA: Sage.

Webb, E. J., Campbell, D. T., Schwartz, R. D., Sechrest, L., & Grove, J. B. (1981). *Nonreactive measures in the social sciences* (2nd ed.). Boston: Houghton Mifflin.

Whyte, W. F., & Whyte, K. K. (1984). *Learning from the field: A guide from experience.* Beverly Hills, CA: Sage.

5 Behavioral Mapping and Trace Measures

Behavioral Mapping

A special application of observational techniques, *behaviorial mapping* is concerned specifically with people's behavior in their environments. In which settings do people spend their time, and how do these settings affect their behavior? Behavioral mapping involves an actual chart or plan of an area on which people's locations and activities are indicated. The mapping is done either at the time of the observation or later from records made in narrative form. This technique has practical relevance for urban design, transportation planning, geography, park management, and other fields concerned with people's locations and movements.

In a study of public housing in Cleveland, Ohio, researchers recruited tenants to serve as observers. The tenants were given special training so that they would record the kinds of behaviors observed in the project area, who was engaged in them, and how long the behaviors lasted. The observers identified 67 distinct behavior settings in and around the project. *A behavior setting* is a geographical location linked to customary patterns of behavior occurring over regular time periods. Examples of behavior settings include community meetings, laundry room activities, playground activities, and the use of the streets and sidewalks. In the number and range of behavior settings, the public housing project actually provided a richer and more varied environment than was available in nearby small towns or poor urban neighborhoods. However, most of these settings in the public housing project were outside the control of the tenants. The housing project consisted of alien, bureaucratic, and institutional spaces controlled by an impersonal management or by no one. It was this lack of control over space, rather than the number and variety of behavior settings, that differentiated the public housing project from other neighborhoods occupied by poor people. In addition to recommending specific architectural/environmental changes to improve the housing project, the research team also urged that more encouragement be given to local decision making and leadership (Bechtel, 1971).

A behavioral map, as indicated earlier, is a chart of people's locations in space. It is an empirical document that describes what behaviors actually occur rather than what was planned for the space (Ittelson, Proshansky, & Rivlin, 1970.) A neighborhood street may be more than a transit corridor. It may also be a place where people congregate, where children play, and in some areas, where drugs are traded.

In constructing a behavioral map, the observer can record only those items that are readily apparent, such as approximate age, sex, whether the person is alone or in a group, and what he or she is doing. There are various ways to record people's locations in space, including time-lapse photography, videotape, and prepared diagrams of an area filled out by an observer. Whatever method is used, the researcher must determine its effects upon the behavior of the respondents. Ittelson, Rivlin, and Proshansky (1976) found a high correspondence between their own observations of how hospital patients spent their time and the patient's own written estimates. It may also be possible to compare people's behavior when they know they are being watched and when they are unaware.

Use of this technique requires observers who are trained to make observations in a standardized and reliable manner. They must be familiar with the terms on the observation sheet. This will require practice, feedback, more practice, and more feedback. Testing for the reliability of the categories is essential. Some researchers use 90 percent agreement among observers in each category as a criterion. Where the observers cannot agree at this level on a specific behavior (not overall, but on a specific behavior), that behavior may be dropped from the data analysis.

Behavioral mapping is often used with children for whom interviews or questionnaires are less appropriate. Teams of psychologists in Finland and the Netherlands charted the movement of preschool children with particular reference to street crossings (Nummenmaa & Syvänen, 1974; Van der Molen, Kerkhof, & Jong, 1983). In both nations, the observations were used to develop traffic education programs for children. The educational program in Finland was distributed throughout the school system and has been in use for over a decade. During this period, road casualties among children dropped 60–70 percent (Nummenmaa & Syvänen, 1974).

Behavioral maps can be place-centered or person-centered. A *place-centered* map shows how people arrange themselves within a particular location. A *person-centered* map shows people's movements and activities over a specified period of time. To construct a place-centered map, observers station themselves to watch the action in a particular location. They record the location and activities of the people in the area on prepared diagrams. A person-centered map begins with a designated number of individuals whose activities are charted throughout the day. Because the observations are very detailed, a person-centered map usually involves only a few individuals. One researcher who wanted to compare the residents of a convalescent home with participants in a day treatment center (nonresidential) selected eight people from each setting who were matched in age, health, educational level, and other characteristics and followed their activities throughout the day. She wanted to find out how much time the persons in each setting spent in social activities, how long they slept, whether they did craft work, and other details. On the basis of her observations, she was able to conclude that residents of the day treatment program were more active than were those in the convalescent home (Quan, 1978).

The choice of mapping procedure will depend on the researcher's goals. If the objective is to assess a particular location, such as the suitability of a room, building, or park, place-centered methods are preferable. If the goal is to learn about a group or individuals, such as the situation of elderly hospital patients, the observer will probably choose a person-centered approach.

Place-Centered Maps

The first task is to draw a diagram showing all those architectural/environmental features that affect people's behavior. This should be based on your own measurements rather than the architect's blueprints, which are probably out of date and do not contain the portable features of the environment, such as chairs, tables, bulletin boards, wall clocks, and information signs. You can use existing floor plans or charts as a guide, but be sure to check the measurements yourself. The map can be drawn in ink and copied. Architectural features can be drawn schematically or symbolically as circles, squares, and triangles. A stationery store will have Plexiglas stencils used by designers for tracing the outlines of chairs, tables, desks, and so on. You do not have to be a professional drafter to construct an outline map.

The next step is to list the behaviors to be recorded. Special scoring symbols are then used for each behavior, (i.e., Si—sitting, St—standing, Wa—walking, etc.). The list is compiled through casual observation of the setting. Categories should cover 80–90 percent of the behaviors found in the setting. Special symbols such as 01 and 02 can be used for "other behaviors," which are described in detail at the bottom of the map. The setting is then sampled at predetermined times and the behaviors are recorded.

William Whyte (1980) used place-centered maps to record plaza use in major cities. The mapping was done while walking through the plaza. Whyte found that he could map the location of every person sitting in the plaza, whether male or female, alone or with others, in about 5 minutes, which was slightly more time than a simple head count would require. The procedure

Behavioral maps. Place-centered maps were used to study space utilization in a university library. Students preferred to sit at the ends of the tables, since this minimized eye contact with others.

was used to determine patterns of plaza usage according to day of the week, weather, season, and the presence of various environmental amenities such as trees, tables, chairs, and food service. Whyte observed interesting changes in himself as he mapped a place. Mapping seemed to give him a proprietary interest; it became *his* place, and the people were *his* people.

The research team in a mental hospital ward found that the categories listed in Table 5–1 included most of the behaviors they observed in the dayroom. When an unusual event occurred on the ward, such as a group sing led by the recreation therapist that took place about twice a month, this was scored in the "other" category and described at the bottom of the record sheet. Such events occurred so infrequently on the ward that the "other" category was not often used. The clearest finding from the observations was the minimal level of activity in the dayroom. Most of the occupants were sitting and doing nothing most of the time.

The reliability of the notation system should be checked. Some behaviors are difficult to categorize. On the hospital ward mentioned earlier, the researchers had a hard time distinguishing between patients sitting and those

TABLE 5-1. Activity List for a Mental Hospital Dayroom

Activity	*Scoring symbol*
Reading	R
Playing cards	C
Watching TV	TV
Listening to radio	Rad
Talking to other patient	T-P
Talking to nurse	T-N
Jigsaw puzzle, games	G
Sitting, doing nothing	S
Standing, doing nothing	St
Sleeping	Sp
In transit (walking)	W
Other (describe)	O_1, O_2

watching television. Observers disagreed among themselves so often in making this determination that sitting (S) and watching television (TV) were combined into a single category. There is no point in including categories of behavior that cannot be reliably observed. Some of the more common methods for estimating scoring reliability are a percentage agreement score or a correlation coefficient (see Chapter 18). When observers disagree frequently as to the category in which a behavior belongs, further refinement of the scoring system and/or training of the observers is warranted.

The map should cover all of the possible times when the area is being used. Some of the most interesting information from behavioral maps is recorded when "nothing is happening." One design team studying a school yard was struck by the number of times during the week when the yard was unoccupied. This led the designers to search for ways to increase community use of the school yard outside of school hours by including a picnic area, amphitheater, and other facilities in the renovation plan (Hester, 1975).

The observation sheet used by the research team is shown in Figure 5-1. The actual usage patterns of the school yard are shown in Figure 5-2.

A group of researchers studying desegregation mapped seating patterns in college classrooms. They hypothesized that if race were not a factor in seating preference, then the number of black students sitting adjacent to white students should be similar to what would be expected in random seating. An observer visited classrooms and mapped seating according to the student's ethnic status. People of different races were more likely to be sitting next to one another in a college known to be less prejudiced than at a college that

Date Hour Weather Observer

	Pre-School		1-8 Grade		9-12 Grade		College		Adult		Elderly		Social Activity*	Setting for Interaction**
	F	M	F	M	F	M	F	M			F	M		
a. Walking														
b. Sitting														
c. Working														
d. Stop to talk														
e. Neighborhood meeting														
f. Active recreation														
g.														
h.														
i.														
j.														
k.														
l.														
m.														
n.														
o.														
p.														
q.														
r. Commercial														
s. Waiting for transportation														
t. Art														
u. Fantasy play														
v. Construction														
w. Role play														
x.														
y.														

*Record the Interaction Process associated with each activity and the number of people interacting in that manner. Use the key: P = Private, I = Impersonal, C1 = Cooperation, C2 = Competition, C3 = Conflict, A = Accommodation.
**Indicate the setting in which each activity takes place and the numbers of people, particularly in that setting.

FIG. 5-1. Observational sheet used in the behavioral mapping of a school yard.

FRED OLDS SCHOOL
INTENSITY OF ACTIVITY
TYPICAL WEEKDAY

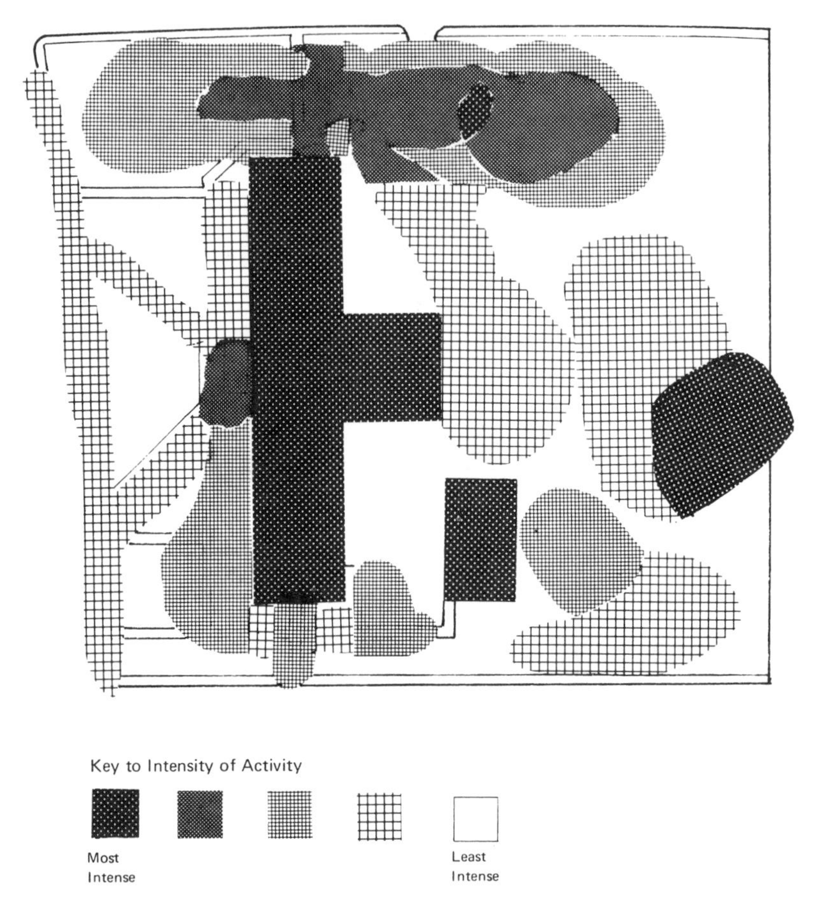

FIG. 5-2. Composite map of interaction throughout the week, made by combining observational records over time.

had been described previously as more prejudiced (Campbell, Kruskal, & Wallace, 1966).

In making your own observations, do not space the sessions too closely together on the same day. Otherwise, you are likely to see the same people doing the same things. To cover all the morning hours, 8 A.M. observations could be made on Monday, 9 A.M. sessions on Wednesday, 10 A.M. observations on Tuesday, and so on.

You will be pleasantly surprised how quickly place-centered observations can be done even for a comparatively large area. A place-centered map deals with instantaneous cross sections of behavior. Generally you will want to record what people are doing when you first see them. What they do afterward is *not* recorded. This will yield a series of behavioral maps at representative intervals throughout the day. Individual maps can later be combined into a single *composite map* showing location, density, and usage for all time periods (see Figure 5-2).

Person-Centered Maps

Person-centered maps are especially useful in a comparative study of environments. Matching the subjects allows the researchers to interpret differences as being due to the settings rather than the characteristics of the subjects. Judy Quan, in the study mentioned earlier, found that residents in the day treatment center were more active than residents of the convalescent home. Since she had matched these individuals beforehand on age, health, and other factors, it seemed logical to attribute the difference to the two environments rather than the characteristics of the two groups of residents.

In constructing a person-centered map, the first task is to identify the sample and obtain subjects' cooperation. At the outset, the subjects should realize the implications of being followed around throughout the day. They must be assured that anything seen or heard will be kept confidential. For recording activities, it is helpful to use a coded notation system that does not involve actual names or details of conversations.

No matter how skillful you are in gaining cooperation, a period of adjustment will be necessary to accustom the person to your presence. Observations may be omitted during this time. Not writing things down during the early sessions will shorten the adjustment period. You might try observing the person in stages—1 hour the first day, 2 hours after that, a morning observation next, and finally an 8-hour session.

Since there are only a few people to be observed, you can expect to be bored much of the time. While you cannot avoid being drawn into conversation occasionally, becoming the person's best friend will detract from the validity of the procedure. Thus, the notation "talking to the observer" should appear infrequently on your scoring sheet.

Person-centered maps are also used in observational studies of children (e.g., in comparing play patterns of children of different ages, of the two sexes, or in different settings). At first, the observer's presence will be noted, but soon the children will habituate and return to their usual patterns of behavior.

Observations can be continuous or periodic. Continuous observations involve following an individual over a period of time (e.g., an 8-hour work shift or an entire school day). Periodic observations involve observing the same individual at intervals throughout the day. This will include a *time sampling procedure* in which the observer prepares a list of specific times when the person's activities will be charted, such as 9:18 A.M., 10:04, and 2:35 P.M. At the designated time, the observer notes the person's location and activity. Time sampling does not require the observer to watch a single subject all the time. The researcher can schedule other activities, including observational sessions with other subjects, during the intervals.

With people whose schedules and activities are regular and predictable, little is gained by extending the observations indefinitely. At some point, it should be more profitable to move to the next person in your sample. Eventually you may want to return to the first person for a second series of observations, but this is optional. The decision as to how many people to include depends more on economics and interpersonal relations than on theory. How long will subjects permit your observation? How valuable are the additional data? If there is an impressive consistency in people's movements and activities, a small representative sample is probably sufficient.

After you have collected a series of person-centered maps, they can be converted to spatial form. For adults this may be accomplished by using an ordinary city map. The person's travels each day are recorded as single lines, which become thicker with more frequent travel. The maps can also be summarized in tables showing the percentage of time spent in various locations and activities (e.g., for children on a playground, the amount of time or number of occasions of swing use, sandbox play, climbing, or games).

BOX 5-1. Behavioral Mapping of Penguins in an Oceanarium

Zoos and animal parks are wonderful places for observing animal behavior, human behavior, and human and animal interactions. Sue Joseph obtained permission from Sea World, a large oceanarium in San Diego, California, to conduct observations in the penguin exhibit. She wanted to learn how use of the various parts of the exhibit changed over the breeding cycle. Her first step was to develop categories of penguin behavior that could be reliably scored by an observer stationed 8–10 feet away. She and her assistants used the categories over a 6-month period in observing the penguin exhibit. The categories on the observation sheet (Figure 5-3) were abbreviated and not always meaningful to an outsider. Longer descriptions of each category were given to her assistants during the training sessions in which the reliabilty of the observers was determined. For example, "ecstatic" refers to a display in which the penguin slowly stretches its head and bill straight upwards while slowly flapping its wings and emitting a staccato vocalization. "Bill-to-axilla" refers to the bird leaning forward at an angle of approximately 45° with slowly flapping wings while rocking its head from side to side with the tip of the bill directed to the base of one wing (after Sladen, 1958). A map of the observation area was drawn and divided into rectangular grids that could be identified through a coordinate system (e.g., Bird 102 was in section B-6 at the time of the observation). Maps of the observation area were placed at the top of the behavioral categories on each score sheet. This score sheet was used for ten 30-second observations of a single penguin throughout a 5-minute period.

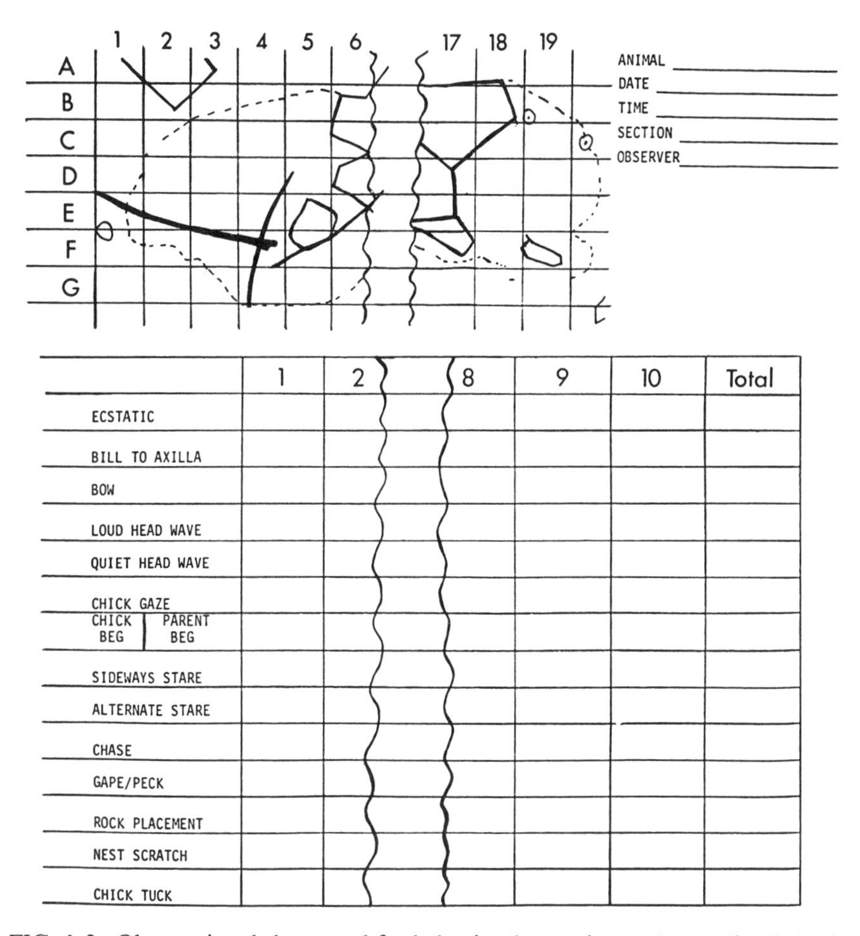

	1	2	8	9	10	Total
ECSTATIC						
BILL TO AXILLA						
BOW						
LOUD HEAD WAVE						
QUIET HEAD WAVE						
CHICK GAZE						
CHICK BEG / PARENT BEG						
SIDEWAYS STARE						
ALTERNATE STARE						
CHASE						
GAPE/PECK						
ROCK PLACEMENT						
NEST SCRATCH						
CHICK TUCK						

FIG. 5-3. Observational sheet used for behavioral mapping and recording behavior of penguins.

Safeguarding the privacy of your respondents is exceedingly important. Check your data carefully to ensure that individuals cannot be identified by name. When it is impossible to prevent identification, as in mapping the activities of a lone school nurse, consider omitting detailed information on his or her activities in any published report.

Limitations

Person-centered mapping is both tiring and intrusive. The observer who follows someone around will have difficulty remaining in the background not only for the person observed but for others encountered. Time sampling alleviates some of the tedium but not the intrusiveness of the procedure.

Behavioral mapping. Several times a day, the location of each penguin in the Sea World Exhibit was indicated by the researcher on prepared maps. (Sea World photo)

Posed picture of behavioral mapping of the penguin exhibit showing the observer with grid diagram on the clipboard and penguins marked with flipper bands. Normally the observations were made from a longer distance away from the birds. (Sea World photo)

Place-centered maps are intrusive in a private location, such as an apartment or personal office; they can work well in public locations. The researcher mapping people's use of a shopping mall may be able to appear as just another customer taking periodic walks through the area.

Combining maps from several individuals observed at various times makes interpretation more difficult. A combined behavioral map for 10 observational sessions in a school yard may show 10 dots at a specific location. The reader does not know whether this refers to one child observed at that location for all 10 sessions or 10 children on one occasion. This confusion of within- and between-individual variation makes it difficult to deal with behavioral maps statistically.

Because of the time and effort involved in their construction, person-centered maps tend to involve only a few individuals. This also makes it difficult to test the data statistically.

Finally, person-centered mapping reveals environmental choices with no information as to why these choices are made. For any practical purpose, the maps must be supplemented with interviews or other data-gathering procedure to find out why certain locations are chosen and others avoided.

Trace Measures

"Moroccans keep out!" North Africa? No. Philadelphia? Yes. And you had better know what it means if you are a local teenager. The message indicates gang activity in the area. This one was a warning to the powerful Moroccan gang to stay out of the territory controlled by its rival. Geographers David Ley and Roman Cybriwsky (1974) analyzed wall writing to chart gang activity in Philadelphia. They found that graffiti are used by teenage gangs to warn outsiders that they are entering someone's turf and must respect local territorial claims.

There are regional and local styles in wall writing. Graffiti in Hispanic neighborhoods in the United States often display a distinctive style of writing (see photograph on p. 74). Code words are frequently employed, such as *M* for marijuana or *13,* since *M* is the thirteenth letter of the alphabet. *Taboo 12-P* means that the turf is controlled by Philadelphia's Twelfth and Poplar Street Gang. Street gangs, lacking legitimate ownership of property, take to the walls to assert their claims to turf. Los Angeles County started a youth gang committee that among other things, keeps track of graffiti that might indicate a turf battle is in the offing. One employee for the committee spent 5 years learning to decipher the graffiti, colors, hand signals, and distinctive manner of speech in gang-plagued areas.

Another type of illegal wall writing is latrinalia, the technical name given to graffiti on public toilet walls. This writing has been studied by anthropologists and others in an effort to learn about cultural values. Latrinalia tend to express attitudes and sentiments that are socially disapproved. Most American toilet walls reveal a preoccupation with sexual experience, excre-

Trace measures: accretion. Graffiti on the side of this East Los Angeles market provides a record of gang activity in the area.

tion, ethnic hostility, and divergent political views. A comparison of latrinalia in Chicago and the Philippines showed far more homosexual writing in Chicago restrooms but about the same amount of heterosexual writing. The American inscriptions were also more likely to be judged humorous, political, and philosophical. The Philippine wall writing had a higher frequency of hostile content and disapproval of sexual ideas (Sechrest & Flores, 1969). The eruption of Mount Vesuvius in 79 A.D. covered the city of Pompeii with a heavy layer of volcanic ash that cooled as it fell to earth. The dense layer of ashes protected the buildings, the walls, and many common household objects for over a thousand years. When the city was excavated, the writing scratched on walls was found to be intact and provided many clues to the day-to-day life of ordinary people.

Using trace measures to study animal behavior, researchers count the number of bird nests to obtain information on density and breeding behavior. There are studies of animal droppings and the gut contents of dead animals (often road kills) to learn what animals eat. A team of researchers in England used trace measures to investigate feeding habits of domestic cats. Through personal contact, all but one of the households in a small village agreed to bag the remains of any animals that the cat caught. These were collected by the researcher on a weekly basis and tabulated. The study showed that small mammals represented almost three-quarters of the prey of these domestic cats. Contrary to what many people believed, birds were always in the minority as prey except during the coldest part of the winter, when mammals stayed below ground. Cats in houses near the middle of the village caught fewer prey than cats living near the edge of the village (Churcher & Lawton, 1989).

Trace measures are the physical remains of interaction after the participants have departed. Webb, Campbell, Schwartz, Sechrest, and Grove (1981) identify two types of physical traces useful in behavioral research—*accretion* and *erosion*. Accretion refers to the buildup of a residue or product of the interaction—something added to the situation. Erosion refers to deterioration or wear and provides an index of usage patterns. With erosion, something is worn down or removed.

Accretion Methods

Graffiti are a form of accretion; litter is another. While the presence of graffiti has been used to chart gang activity, litter has been employed to study usage patterns in public parks. Garbage, another accretion measure, has been used in the study of human nutrition to determine what and how much people are eating and how much useful food is discarded. Sawyer (1961) made an estimate of people's drinking habits based on the number and kind of liquor bottles found in trash cans.

Erosion Methods

Certain patterns of usage create certain types of erosion or wear. Deterioration of campus lawns and the extent of footprints in the snow reveal informal pathways. Astute university admininstrators delay installing walkways until after these informal pathways have been located. Such pathways can be used to assess the social relationship between people in different buildings. Footprints do not provide an infallible measure of contact, but this clue may be worth following up if there is a significant number of footprints in the snow going from one dormitory to another and none between that dormitory and another equally close. According to Webb et al. (1981), the floor tiles around the hatching chick exhibit at Chicago's Museum of Science and Industry must be replaced every 6 weeks, while tiles in other parts of the museum last for years. Tile wear, they suggest, is a good index of the relative popularity of exhibits. In the hallways on our campus, the floor tiles with the greatest number of cigarette burns are those directly outside classrooms where students congregate between lectures. Although it is relatively easy to record waiting patterns through systematic observations, cigarette burns increase confidence in the findings.

Trace measures: erosion. Informal paths shed light on the interaction networks among these apartment dwellers.

Limitations

Apart from archeology, trace measures have not found wide acceptance in behavioral research. Most social scientists prefer to study live interaction rather than carpets and lawns. Analyzing garbage and litter can be a dirty business that involves a serious breach of privacy. For example, park employees in Michigan objected when they were ordered to dig through garbage barrels as part of a research project on throwaway bottles (Associated Press, 1978).

The information provided by trace measures can be misleading. Deterioration of certain library books, for example, may reflect the activities of a single destructive individual rather than a large general circulation. The absence of graffiti in a restroom may reveal more about the cleaning policies of the janitors than about the social attitudes of the people using the restroom. The wear on the floor tiles around drinking fountains will be due to a combination of usage and dripping water.

Summary

A behavioral map is a record of people's behavior in space. Behavioral maps can be place-centered or person-centered. A place-centered map shows how people arrange themselves in a particular setting. A person-centered map involves a record of a person's movements and activities over time.

Limitations: Behavioral mapping can be both tiring and intrusive. Further research will be needed to explain the behaviors observed.

Trace measures are the physical remains of interaction after the people have departed. Accretion refers to the buildup of a residue or product of the interaction. It is something added to the situation. Erosion refers to deterioration or wear and provides an index of usage patterns. With erosion, something is worn down or removed.

Limitations: Trace measures are necessarily indirect and may be the result of extraneous factors.

References

Associated Press. (1978, August 31). High priority research stinks, garbage-picking workers say. *Sacramento Bee,* p. B5.

Bechtel, R. (1971). *Arrowhead: Final recommendations.* Kansas City: Environmental Research and Development Foundation.

Campbell, D. T., Kruskal, W. H. & Wallace, W. P. (1966). Seating aggregation as an index of attitude. *Sociometry, 29,* 1–15.

Churcher, P. B., & Lawton, J. H. (1989, July). Beware of well-fed felines. *Natural History,* pp. 40–47.

Hester, R. T. (1975). *Neighborhood space.* Stroudsburg, PA: Dowden, Hutchinson, & Ross.

Ittelson, W. H., Proshansky, H. M., & Rivlin, L. G. (1970). The environmental psychology of the psychiatric ward. In H. M. Proshansky, W. H. Ittelson, & L. G. Rivlin (Eds.), *Environmental psychology* (pp. 27–36). New York: Holt, Rinehart and Winston.

Ittelson, W. H., Rivlin, L. G., & Proshansky, H. M. (1976). The use of behavioral maps in environmental psychology. In H. M. Proshansky, W. H. Ittelson, & L. G. Rivlin (Eds.), *Environmental psychology* (2nd ed.) (pp. 340–350). New York: Holt, Rinehart and Winston

Ley, D., & Cybriwsky, R. (1974). Urban graffiti as territorial markers. *Annals of the Association of American Geographers, 64.* 491–505.

Nummenmaa, T., & Syvänen, M. (1974). Teaching road safety for children in the age range of 5–7 years. *Pedagogica Europaea, 9*(1),

Quan, J. (1978). *Social interaction of Chinese elderly in a skilled nursing home and day health center.* Unpublished manuscript.

Sechrest, L., & Flores, L. (1969). Homosexuality in the Philippines and the United States: The handwriting on the wall. *Journal of Social Psychology, 79,* 3–12.

Sladen, J.J.L. (1958). The pygoscelid penguins, parts 1 and 2. *Scientific Reports of the Falkland Islands Dependency Survey, 17.* London.

Van der Molen, H. H., Kerkof, J. H., and Jong, A. M. (1983). Training observers to follow children and score their road-crossing behavior. *Ergonomics, 26,* 535–553.

Webb, E. J., Campbell, D. T., Schwarz, R. D., Sechrest, L., & Grove, J. B. (1981). *Nonreactive measures in the social sciences* (2nd ed.). Boston: Houghton Mifflin.

Whyte, W. H. (1980). *The social life of small urban spaces.* Washington, DC: The Conservation Foundation.

Further Reading

Castleman, C. (1982). *Getting up: Subway graffiti in New York.* Cambridge, MA: MIT Press.

Webb, E. J., Campbell, D. T., Schwartz, R. D., Sechrest, L., & Grove, J. B. (1981). *Nonreactive measures in the social sciences* (2nd ed.). Boston: Houghton Mifflin.

Whyte, W. H. (1980). *The social life of small urban spaces.* Washington DC: The Conservation Foundation.

Wicker, A. W. (1987). Behavior settings reconsidered. In D. Stokols, & I. Altman (Eds.), *Handbook of environmental psychology* (pp. 613–649). New York: Wiley.

Willems, E. P. (1977). Behavioral ecology. In D. Stokols (Ed.), *Perspectives on environment and behavior.* New York: Plenum Press.

III
Designing Experiments

6 Experimentation

Toxic exposure is receiving considerable public attention as a health hazard. Many industrial workers are exposed to solvents and other chemicals, and there have been reports that these chemicals can damage the central nervous system (Iregren, Åkerstedt, Olson, & Gamberale, 1986). The damage almost always occurs among workers exposed to a variety of solvents and there are indications that some of the problems involve interactions with alcohol use. Researchers associated with the Swedish National Board of Occupational Safety and Health decided to use laboratory experimentation to sort out the competing explanations as to whether the solvent toluene produced mental or physical problems and whether it interacted with alcohol use.

The details of the procedure and the apparatus are described in the original article (Iregren, Åkerstedt, Olson, and Gamberale, 1986). Here we are interested in the way laboratory experimentation was used by the researchers to answer difficult and important questions. A team of four researchers developed a battery of psychological and physiological tests. The psychological tests included two measures of reaction time (how quickly people respond in a designated task), color–word vigilance (a task that creates competition between two ways of perceiving stimuli), and a memory task involving letters and numbers.

The experiment took place in a special chamber in the research unit. The subjects sat in the chamber and breathed air containing calibrated amounts of toluene. The 12 subjects were healthy volunteers who were carefully screened, provided with detailed information beforehand about the experiment, and closely monitored during and after the experimental sessions. The level of toluene in the experiment corresponded to the accepted Swedish Industrial Standard.

Each subject participated in a training session, followed by four experimental sessions. The subjects were tested individually in the chamber and each session required almost a full day. The order of sessions was counterbalanced to control for order effects (i.e., the effects of the sequence of tests). The four sessions included one control session with no solvent or alcohol, one session with the solvent alone, one session in which the subject drank the alcohol mixed with orange juice, and one with a combination of solvent and alcohol. The counterbalanced order of sessions is shown in Table 6-1.

The psychological tests were administered through a computer terminal inside the chamber. Physiological recording was done continuously throughout the session using small electrodes placed on the subject's chest and head. During each session the subjects also filled out a mood checklist and a symptom scale.

With all the different tests and sessions, the experiment took several months to complete. Data analysis, involving numerous measures and comparisons, was also complex and time-consuming. However, after the experiment was concluded, the researchers could state with some confidence what they had found and what it meant. They had not discovered Truth, but smaller truths about the effects of a common industrial solvent, about people's physiology, mental state, and behavior. To remove the suspense (you have probably read more than you wanted to about toluene), here is a capsule summary of the findings:

Toluene exposure at an "acceptable level" produced headaches and local irritation, and a brief lowering of heart rate, but did not reduce the performance on psychological tests.

Alcohol affected both mood and two psychological measures (slower reaction time and worse performance on color–word vigilance).

No interaction was found between alcohol and toluene exposure at these levels. That is, the two in combination did not produce different results than would be expected from either alone.

TABLE 6-1. Experimental design for study of solvent and alcohol effects

	Treatment Condition			
Session	Group 1 ($n = 3$)	Group 2 ($n = 3$)	Group 3 ($n = 3$)	Group 4 ($n = 3$)
1	control	solvent	alcohol	solvent & alcohol
2	solvent	control	solvent & alcohol	alcohol
3	alcohol	solvent & alcohol	control	solvent
4	solvent & alcohol	alcohol	solvent	control

These conclusions are valid only for the exposure levels and measures (physiological and psychological) used in the study.

The findings were published in a technical journal and transmitted to the public health agency that had supported the research. They were also given directly to the trade unions of industrial workers exposed to toluene.

The Special Nature of the Experiment

This study illustrates how the experimental method is used to separate alternative explanations and test them. Among the various research methods, the experiment is particularly suitable for tracing cause-and-effect relationships. Through observation one finds many things that occur together, but observation alone cannot determine whether one thing is the cause of another. All other alternative explanations must be ruled out. A classic example is the relationship between cigarette smoking and lung cancer. Doubts about the contribution of cigarettes to lung cancer were removed when experiments on animals showed that the substances in cigarettes were carcinogenic (cancer causing). Other environmental factors, such as air pollution, may also affect the relationship. The experiment enables each of these factors to be tested in a systematic way. An artificial situation is created in which events that generally go together may be pulled apart.

In his book *Zen and the Art of Motorcycle Maintenance,* Robert Pirsig (1974) describes the purpose of the experimental method as making sure that Nature has not misled you into believing you know something you actually don't know. As Pirsig describes it, the power of the scientific method, and experimentation in particular, outweighs the elaborate preparation and formalities required. This does not mean that an experiment always provides predicted results. Nature can jealously guard her secrets and you can end up learning a fraction of what you expected to in addition to much more than you ever imagined. An experiment fails only when it does not adequately test the hypothesis. If an experiment is properly conducted, even results that do not support the hypothesis may be useful and important.

Variables

A *variable* is any characteristic or quality that differs in degree or kind and can be measured. Examples of variables are height, hair color, sex, running speed, income, IQ, education, social class, and political party. Variables have *values* or *levels.* Some variables are *continuous* in that they possess a sliding scale of value, varying in quantitative fashion. Age is a continuous variable, ranging from zero to over 100 years. A person's age may fall anywhere along that dimension. Other variables are *categorical:* Mutually exclusive categories of value, instead of continuous dimensions, are used. The levels tend to be more qualitative than quantitative. Political party membership is a variable with categorical values, as is sex. A person belongs to one or another party and is usually either male or female. Some variables may be treated as either continuous or categorical. If hair color is measured on a scale of very light to very dark, the variable will have continuous values. On the other hand, if we use mutually exclusive categories of blond, brunette, and redhead, hair color would be considered a categorical variable with three levels.

Independent and Dependent Variables

The variable that is systematically altered by the experimenter is called the *independent variable.* In an experiment, the treatment is the independent variable. The variable that we presume it will affect is called the *dependent variable.* Synonyms for the independent variable are experimental or predictor variable, factor or treatment. The dependent variable is sometimes referred to as the *outcome, response,* or *criterion variable.*

Experimental Control

The purpose of an experiment is to determine the effect of the independent variable upon the dependent variable. All other influential variables must be either eliminated or their effect controlled in some way to ensure that they are not the cause of a change in the dependent variable. The experimenter should avoid *confounding* or confusing the effect of the independent variable with that of other variables. These other variables to be either eliminated or controlled are termed *extraneous.* They are neither the independent nor the dependent variables of the study. An extraneous variable in one experiment may become an independent or dependent variable in another. Extraneous variables are sources of error, since they mask or cloud the observable effects of the independent variable. For example, in an algebra experiment, if the spaced-study group were more intelligent than the massed-study group, the independent variable would be confounded with the extraneous variable of intelligence. We would not know whether a difference between the groups resulted from the type of preparation or a difference in intelligence. Extraneous variables are associated with the subjects, the experimenter, the setting in which the experiment occurs, and the experimental procedure itself.

Subject Characteristics

The participants in an experiment are called *subjects.* The qualities of the subject are likely to influence the experimental outcome. These include such characteristics as age, education, socioeconomic background, and personality, as well as temporary conditions such as fatigue, nervousness, or preoccupation with other matters. Participants' expectations and beliefs about the experiment may also influence the results.

Solutions:

1. Eliminate or reduce extraneous variables when possible; for example, reduce fatigue by using rested subjects.
2. Make the extraneous variables as similar as possible for all treatment conditions
   a. *Matching groups*—each treatment group shows similar levels of subject characteristics. They are matched by sex, religion, age, socioeconomic background, intelligence, and so on.
   b. *Random assignment*—subjects are randomly assigned to treatment groups. The assumption is that the extraneous variables will be randomly distributed, and thus roughly the same for all treatment groups.

This ensures that extraneous variables will be roughly the same for all conditions. See Chapter 16 for descriptions of sampling techniques and random selection and assignment procedures.

Experimenter Characteristics

Experimenters, like subjects, possess a variety of characteristics and expectations that might influence the outcome of an experiment. These must be eliminated or kept the same for all subjects.

Solutions

1. Train experimenters in order to reduce extraneous influences.
2. Use the same experimenter for all subjects.
3. If more than one experimenter is involved, have each run an equal number of subjects in each treatment condition, *or*
4. Randomly assign experimenters to subjects.

The Setting

Qualities of the setting in which the experiment takes place—the functioning of the equipment, time of day, and other situational variables—are also potential sources of error.

Solutions

1. Run all subjects about the same time of day.
2. Keep the laboratory conditions (lighting, temperature, noise, etc.) as similar as possible for all subjects (except for the levels of the experimental variable).

3. Check equipment frequently to make sure that it is functioning properly.

The Procedure

The groups treated with or exposed to different levels of the independent variable are called the *experimental groups*. In addition, many experiments use a *control group*. The control group is deliberately set up to be influenced by all of the variables affecting the experiments groups *except* the independent variable.

Example:
A researcher is interested in the effects of a drug on a rat's performance in a maze-running task. Rats are randomly assigned to either the experimental or control conditions.

Subject groups	*Treatment condition*	*Measure*
Experimental	Independent variable (drug injection)	Dependent variable (maze-running time)
Control	Same as the experimental group *except* for the independent variable (salt water instead of drug injection)	Dependent variable (maze-running time)

If changes in the experimental group were due to the experience of being handled or receiving a shot (extraneous variables), the control groups should also show them. The effect of the drug is assessed by looking at the difference beteween the two groups on the dependent variable (maze-running time). A difference is then attributable to the drug.

Sometimes subject performance is measured before and after treatment; thus, the subjects serve as their own controls. A problem, however, is that change might occur as a result of practice, fatigue, or some other occurrence between trials. Thus, a separate control group is often more appropriate.

Types of Experiments

True Experiments

In a true experiment, three principles must hold: (1) at least two groups or conditions are compared, (2) the researcher has control of or can predict and evaluate the experimental treatment, and (3) subjects are randomly assigned to treatment groups. True experiments may be conducted in the laboratory or in the field—classrooms, homes, public streets, and so on. The only difference between laboratory and field experiments is their location.

Laboratory experiments are probably the most common, since they are done in a specially created setting in which the experimenter is able to control a wide variety of extraneous variables. Temperature, lighting, and other aspects of the environment are easily held constant, while distractions such as noise and other activity are eliminated or reduced. In the laboratory, conditions can be created and changed according to schedule and in a manner desired by the researcher. As a convenience to the experimenter, the laboratory is hard to beat. Apparatus is often used in laboratory experiments (see Chapter 14).

Some researchers prefer to do experiments in the outside environment. In a *field experiment* the researcher manipulates the independent variable, but does so in the field (the outside environment) instead of in the laboratory. The design of the field experiment is like that of a laboratory experiment, except for the setting. Field experiments tend to be more difficult to do and interpret than laboratory experiments because there are so many uncontrolled factors that affect behavior: economic conditions, climate, number of people around, and myriad other environmental conditions. Their major advantage is their naturalness compared to the presumed artificiality of the laboratory. Field experiments often possess more external validity than laboratory experiments. That is, the results may be more readily generalized to real-world settings. Experimental psychologists have studied the spatial imagery of taxi drivers (how they learn to find their way around the city), how police are trained to visually process information at the scene of a crime, and the ability of headwaiters to remember customers' names. Henry Bennett (1983) investigated the ability of waiters to remember complicated food and beverage orders. In one study, Bennett and three friends entered a restaurant, sat at a table, and ordered a full-course meal. After the waiter had left the table the four friends switched places. This was done to see if the waiter associated orders with place or with person. Bennett's work qualifies as a field experiment. It was done in a natural setting, and the experimenter deliberately changed a single factor (locations at the table) in a systematic way to determine its effect on the waiter's recall. Had the research not been so time-consuming and expensive (four full-course meals for each experimental session) Bennett could have expanded the study to compare tables with 4, 6, 8, and 10 customers. He also would have liked to compare customer orders given in a clockwise pattern with those given from random locations at the table.

Actually, Bennett did continue his investigation to include these factors, but instead of doing further field experiments with food orders, he undertook a modified simulation (see Chapter 7). He tape recorded complicated beverage orders from tables containing 7, 11, and 15 individuals and compared the ability of cocktail waitresses and college students to match the orders with table locations. He also interviewed waitresses on the methods of recall they used. This is a good illustration of a multimethod approach to a problem that has traditionally been studied only in the laboratory.

Certain topics are more appropriately studied in the field than in the laboratory. For example, responses to litter and reactions in dentists' waiting rooms are best assessed in the real world. Field studies allow the researcher to go beyond verbal replies to questionnaire or interview items (How often do you litter? Would you pick up an empty beer can lying on the road?), and they avoid the artificiality and distortion of the laboratory. Following the hypothesis that "litter begets litter," it has been found that areas deliberately littered by the experimenter will be more littered by other people than those areas kept neat and clean (Finnie, 1973).

As a compromise between the rigorous control of the laboratory and the freedom of natural observation, the field experiment loses some of the advantages of both techniques. Typically there is less control over field conditions than is possible in a laboratory, and the requirements of the experiment limit the naturalness of the conditions studied.

Quasi-experiments

The distinguishing characteristic of a *quasi-experiment,* also known as a *natural experiment,* is that the third condition for a true experiment *(random assignment)* is not met. These are studies in which it has not been possible or feasible to randomly assign subjects to treatment groups or conditions. The "treatment" may be a natural event, a new law or program, or something else outside the researcher's control. Either a before-and-after design, or a comparison of an exposed group with an unaffected one, may be appropriate. Examples of quasi-experiments are:

Before-and-After Designs

- The rate of highway accidents before and after the setting of a new speed limit.
- Reproduction rates in a species of birds before and after El Niño, an occasional warming of the ocean that takes place off the coast of South America.
- Number of psychiatric hospital admissions in a community before and after a natural disaster.

Treatment versus Control Designs

- School achievement in two districts with different student-teacher ratios.
- Sales of sleeping pills in two similar communities, one of which has been struck by a tornado.
- Achievement motivation in first-born children compared with later-born children.

In these instances the researchers do not have direct control over the application of the independent variable, although its effect can be evaluated.

These studies gain much in external validity, because they take place in the real world. Unfortunately they lose internal validity. Because subjects are not randomly assigned to conditions, there may have been self-selection or some other form of bias that in turn confounds the results. For example, people living in the town with the lower student–teacher ratio may be wealthier, so that differences in school achievement may in fact be due to home environment rather than to classroom variables.

How to Do an Experiment

Selecting a Topic

Although you may begin with fairly vague ideas, these must be formulated clearly before you design an experiment. Many ideas are suitable for experimentation, but some questions are more important than others. You might

Quasi-experiment. To learn how much distance motorists kept between themselves and a bicyclist riding at different speeds, a bicyclist rode on a straight path along a road with separate 25-mile, 35-mile, and 45-mile speed limits. Another researcher photographed the "shy distance" kept by cars in each speed zone.

ask yourself the grandmother question, "Why is a grown person like yourself concerned with this question?" If you are not sure about your answer, either give the question more careful thought or change topics. You should be able to explain in clear language what you are doing and why you are doing it.

Developing Hypotheses

The experimental question concerns the relationship between two or more variables. This is generally phrased as a *hypothesis,* a testable proposition.

> *Example*
> Question: Do students in desegregrated classes show a change in IQ scores?
> Hypothesis: Students in desegregated classes will show an increase in IQ scores over time.

This hypothesis is stated positively: there will be a change. It is also possible to rephrase the question as a *null hypothesis,* which predicts that there will be no change.

> *Example*
> Question: Do students in desegregated classes show a change in IQ scores?
> Null hypothesis: Students in desegregated classes will show no change in IQ scores.

After the hypothesis has been formulated, you can consider whether an experiment would be helpful and, if so, what type of experiment—laboratory, field, or natural (quasi). At the conclusion of the study, the hypothesis will either be accepted or rejected, depending on the results.

Operational Definitions

The concepts included in the hypothesis must be operationally defined. An *operational definition* involves defining something by the way in which it is measured. Intelligence may be operationally defined as a score on an IQ test. Operational definitions sometimes vary from experiment to experiment. Worker satisfaction may be defined by a number of operations—turnover rate, productivity, expressed satisfaction on a questionnaire, and rate of absenteeism. However, while operational definitions may vary, they are not arbitrary. They must show a logical relationship to the concept under study.

The primary reason for using an operational definition is to specify clearly and precisely what is being measured so that the study may be repeated by someone else using exactly the same procedures and measurements. Using operational definitions reduces the likelihood of misunderstanding.

Designing the Study

The purpose of the experiment is to assess the effects of one or more independent variables on some dependent variable(s). The experimental design must accomplish this and control for the effects of extraneous variables in order to generate conclusive results.

Problem: Are words that refer directly to sense experience (high-imagery words), such as *sun, brown, loud,* and *sweet,* learned more easily than words that do not refer directly to sense experience (low-imagery words), such as *sent, ask, idea,* and *certain*?
Hypothesis: High-imagery words will be better recalled than low-imagery words.
Operational definition: Ease of learning defined as a number of words recalled correctly.
Independent variable: Degree of imagery
Values: High
Low
Dependent variable: Number of words of each type recalled after exposure to them.
Value: Continuous
Extraneous variables: Familiarity of words, word length, ability of subjects, fatigue, distraction in the setting.

Alternative Research Designs

1. All participants are presented with a mixed list of words.

Subject groups	*Treatment condition (Independent variable)*	*Measure (Dependent variable)*
All subjects	Mixed list—high- and low-imagery words	Number of high- and low-imagery words recalled correctly

2. Participants are randomly assigned to group 1 or group 2.

Subject group	*Treatment condition (Independent variable)*	*Measure (Dependent variable)*
Group 1	High-imagery word list	Number of high-imagery words recalled correctly
Group 2	Low-imagery word list	Number of low-imagery words recalled correctly

3. *Counterbalanced* design: Both groups receive the same lists of words, but in different order.

Subject group	*Treatment condition (Independent variable)*	*Measure (Dependent variable)*
Group 1	First list—High-imagery words followed by recall	Number of high-imagery words recalled correctly
	Second list—Low-imagery words followed by recall	Number of low-imagery words recalled correctly
Group 2	(reverse of above)	

The purpose of *counterbalancing* is to balance out effects due to order of presentation. Counterbalancing is one way of dealing with possible confounding of variables. If the low-imagery list were presented last to everyone and subjects did more poorly on recall of low-imagery words, it would not be clear whether the results were due to the low imagery or to some other factor, such as fatigue. If they did better, it might be due to practice; hence the need for counterbalancing.

Subject Selection and Assignment to Treatment Conditions

Follow the procedures outlined in Chapter 16 that describe the various technique for obtaining subject samples. Be sure that the project follows local guidelines for the use of human or animal subjects.

Pilot Testing

Before actually running the experiment, it is essential to run a pilot study in order to be sure that the equipment is working and that directions are easily understood. Pick a small number of persons to use as pilot subjects. Do not use anyone you intend to use in the actual experiment because the pilot test experience might affect their later responses. Think of the pilot study as a last-minute check on the details. Make it as similar to the anticipated experiment as possible. If major problems are encountered, correct them and conduct another pilot test before running the actual experiment. The importance of the pilot test cannot be overemphasized. A preliminary run may save considerable time and effort later.

Running the Experiment

If the pilot test has been successful, the actual experiment should proceed smoothly. Be sure to follow the procedures specified in the design. If you are working with animals, each should be handled in the same manner, except for the specific treatment constituting the independent variable. If you are studying human beings, each person in each group is to be treated in identical fashion—same instructions, same interaction with the experimenter, and so on. All of the considerations mentioned earlier about the control of extraneous variables apply during the experiment.

Be sure to thank participants for their assistance and cooperation. Offer to answer any questions they might have, explain the purpose and uses of the data, and mention when and where the results will be available. See Box 6-1 for a list of steps involved in experimentation.

Limitations

Artificiality is both a major strength and major weakness. Experimental design provides for the separation of variables and enables either control or

---

BOX 6-1. Experimentation Checklist

A. Design considerations, preliminary to performing the experiment
   1. Is the rationale of the research question spelled out and are the hypotheses clearly stated?
   2. Are the independent and dependent variables specified and operationally defined in a logically justifiable manner?
   3. Does the research design control for
      a. Subject error?
      b. Experimenter error?
      c. Environmental error, including apparatus, setting, time of day, and so on?
   4. Is the subject sample representative of the population of concern?
   5. Have you met local requirements concerning the use of human (or animal) subjects?

B. Pilot test (trial experiment)
   1. Are directions clearly understood?
   2. Does the apparatus work correctly?
   3. Is the experimenter performing correctly?

C. Correct any anticipated problems. If they are extensive, do a second pilot test, and upon satisfactory completion proceed with the experiment.

D. On completion, debrief participants and thank them for their cooperation. Answer any questions and explain where and when the results will be available.

---

measurement of extraneous variables. Nature, however, does not operate in accord with this approach. Variables are combined and often interactive. Science is an imposition of a theoretical structure upon the natural world. The structure fits better in some cases than in others. Predictions derived from good experimentation are valid only for settings similar to those in which they were obtained. The more one uses the advantages of the experiment—separating and controlling the effect of variables—the further one moves from the natural world. Subjects in the laboratory may behave differently from people (or animals) outside. Some problems are not suitable for experimentation either for ethical reasons or for cost.

Summary

An experiment involves the creation of an artificial situation in which events that generally go together are pulled apart. The participants in an experiment are called subjects. The elements or factors included in the study are termed variables. Some variables, like age and income, are continuous in that they range along a continuing scale. A variable whose categories are, at least in theory, separate and exclusive, such as membership in a political party or eye color, are termed categorical.

Independent variables are those that are systematically altered by the experimenter. Those items that are affected by the experimental treatment are the dependent variables. Those variables that are not eliminated or controlled are termed extraneous. The most common extraneous variables are those associated with the subjects, the experimenter, the setting, and the experimental procedure.

The experimental groups consist of those subjects exposed to the independent variable. Other subjects not exposed to the independent variable may be included in the study as a control group.

In a true experiment, subjects are randomly assigned to treatment conditions. Laboratory experiments permit control of a wide variety of extraneous variables. Experimental control is more difficult in the field. However, field experiments often have more external validity than laboratory ones. In a quasi- or natural experiment, subjects are not randomly assigned to conditions, but, rather, studied as the conditions naturally occur. Quasi-experiments take advantage of situations that arise as a result of social policy or natural events.

A hypothesis is a testable proposition. The concepts included in the hypothesis must be operationally defined. An operational definition defines something by the way it is measured.

Limitations: While allowing the researcher an impressive amount of control over variables, the experiment in behavioral research is limited chiefly by its artificiality. Conditions in the laboratory may be different from those in the outside world, and people will behave differently. Some behavioral issues are not suitable for experimentation for ethical reasons or cost.

References

Bennett, H. L. (1983). Remembering drink orders: The memory skills of cocktail waitresses. *Human Learning, 2,* 157–169.

Finnie, W. C. (1973). Field experiment in litter control. *Environment and Behavior, 5,* 123–133.

Iregren A., Åkerstedt, T., Olson, B. A., & Gamberale, F. (1986). Experimental exposure to toluene in combination with ethanol intake. *Scandinavian Journal of Work and Environmental Health, 12,* 128–136.

Pirsig, R. (1974). *Zen and the art of motorcycle maintenance.* New York: Morrow.

Further Reading

Cook, T. D., & Campbell, D. T. (1979). *Quasi-experimentation: Design and analysis issues for field settings.* Chicago: Rand McNally.

Keppel, G., & Saufley, W. H., Jr. (1980). *Introduction to design and analysis: A student's handbook.* San Francisco: W. H. Freeman.

McGuigan, F. J. (1990). *Experimental psychology: Methods of research* (5th ed.). Englewood Cliffs, NJ: Prentice-Hall.

Solso, R. L., & Johnson, H. H. (1989). *An introduction to experimental design in psychology: A case study approach* (4th ed.). New York: Harper & Row.

7 Simulation

What does close confinement in a space capsule under conditions of gravity do to one's personal space? How close is "too close" when a fellow astronaut floats by in a horizontal position? Studying interpersonal distance during actual space flights would be expensive and difficult to control. A solution used by a team of researchers under contract to the American space agency was to employ a cardboard model with astronaut dolls. Earlier research had established that the conversational distance people used for dolls in simulated conversation was similar to the distances real people used in actual encounters.

> In front of you are two male figures and a space station scene. I'm going to ask you to put the men in the space station in a comfortable conversational arrangement with one another. Keep in mind that they are in zero gravity. Every time you put them in position, keep their feet off the floor. When you have put them in a comfortable conversational distance, tell me and I will measure them.

These were the instructions used by psychology student Susan Westfall to study interpersonal distance under conditions of weightlessness. The blond-haired "Ken" doll and the dark-haired "Derek" doll were dressed in specially designed one-piece blue jumpsuits with an American flag on the left sleeve and a black name patch on the left side of the chest (see photo). To simulate zero gravity (weightlessness), the figures were mounted on the ends of 36-inch desk lamp arms that could be rotated 360 degrees. The space station scene was a cardboard panel containing silver pipes, control dials, and other mechanical items found in earlier space vehicles. In a control condition, the doll figures were placed in a living room scene. Westfall found that conversational distance was greater in the simulated space station than in the control condition of the living room. The longest conversational distances occurred when one of the "astronauts" was vertical while the other floated either horizontally or upside down.

To simulate conversational distance at zero gravity, students adjusted the distance between dolls in various spatial arrangements.

What Is a Simulation?

Behavioral simulations are imitations of actual conditions. They are intended to resemble the true situation in many of its functional characteristics without being mistaken for it. None of the participants mistook the scale model for a true space station, but seeing the scale model and the dolls in various positions helped them to imagine what weightlessness would be like.

There are many ways to imitate natural conditions. Scale models are one way; experimental games are another. Complex decision-making processes are difficult to study in the real world or to create in the laboratory. A solution is the simulation game in which people take on the roles of participants. In a simulation called Air Pollution Exercise (APEX), people are assigned to play the roles of private land developers, industrialists, city and county planners, politicians, and air pollution control officers attempting to cope with the air quality problems in a city. An APEX player learns the difficulties and frustrations in attempting to bring together individuals and groups with diverse interests, attitudes, and goals.

A simulation involves a complex social situation in which many variables are deliberately mixed together with no effort made to untangle them, as distinct from a laboratory experiment in which variables are introduced systematically so that their specific effects can be measured. Once a simulation has begun, such as when people are asked to imagine that they are participants in a city council meeting on air pollution, it develops its own dynamics and proceeds at its own pace. The researcher lacks the control over the session that would be considered necessary in a laboratory experiment. Simulations have many uses in teaching and clinical practice that will not be covered here. For example, asking people who have been divorced to play the part of a former partner may help them to understand the other person's view of the situation. Readers who are interested in nonresearch applications of simulation will find interesting information in the periodical *Simulation and Games,* published by Sage Publications.

Realism is an important aspect of simulation. When people are able to see around the corners of their blindfolds, the experience probably isn't very similar to blindness. If the images on a videotape simulation are fuzzy and indistinct when people in ordinary life see things clearly, then their responses are probably not going to be realistic. The degree of realism in a simulation depends upon the skill of the experimenter in capturing the significant elements of the experience, the quality of the available equipment, and the willingness of the participants to project themselves into the situation. Typically a successful simulation requires good cooperation from the participants. If they are not willing to "pretend," then their responses are not going to be useful in predicting actual behavior.

There are several guidelines that can be used in gauging the realism in a simulation. The results can be compared with the behavior of people in the actual situation. For example, to what degree did people in the artificial jail act in the same way as genuine prisoners and guards? The latter information can be obtained from interviews with people in jail or from reading autobiographies of prisoners and guards. A second method is to question the respondents afterwards about their experiences. Did they feel like jail inmates and did other participants treat them as if they were?

Simulations vary in cost and complexity. They may be as modest as asking students in a class to imagine themselves participants in a UN conference or as complex as constructing a stationary space capsule in the basement of a NASA laboratory. While the mock-up of a space ship and the detailed monitoring of people's responses cost thousands of dollars, an actual flight trial of a space ship would cost tens of millions. Because it depends heavily on the ability of participants to "fill in" omitted items through imagination, a simulation is less expensive than constructing the actual object.

Simulations also tend to be more economical in terms of time. A graduate student in comparative psychology became interested in improving the information signs at the local zoo. At first, she considered making her own new signs and trying them out on the actual cages. She was interested both in different types of signs and in placing them in different locations outside the cage (e.g., at eye level, at the bottom, at the side). It quickly became apparent that five different signs in five locations would mean 25 experimental conditions. While this would be very awkward and time consuming for a field experiment, it would be a natural problem for a simulation. Once the student created the five different signs, she could photograph these in each of the five locations on the cage. Zoo visitors could then be shown a display of color photographs and asked for their preferences. Testing could be done at a booth in the zoo. Simulation would reduce an otherwise unwieldy field experiment with 25 different conditions to manageable proportions.

Physical handicaps and sensory deficits can also be simulated. This avoids the criticism that the blind or deaf are a select group so well adjusted to their handicaps that they don't recognize how their behavior has changed. However, one would not want to draw conclusions about the world of the blind or the deaf solely on the basis of college students wearing blindfolds and earplugs. It would be necessary to interview blind and deaf people too.

A research team composed of both handicapped and nonhandicapped persons tested the University of California, Santa Cruz, campus for wheelchair access. A fleet of wheelchairs had been provided so that everyone, handicapped and nonhandicapped, could experience some of the problems in getting around. For example, many elevator and emergency alarm buttons were too high to be reached from a wheelchair; so were some copy machine coin slots, many drinking fountains, all upper book shelves, and the top drawers of the library card catalog. Maneuvering through the heavy restroom door in bathroom cubicles, including those designed specifically for wheelchairs, proved very difficult. The team made numerous suggestions for solving the problems they had identified during the exercise (*UCSC Review,* 1978).

Simulation procedures like this can come close to being full-fledged experiments. An experiment involves an artificial situation in which events that generally go together may be separated. A simulation involves an artificial situation, such as a space capsule or wheelchair ride, whose components are left together. To the extent that the researcher attempts to separate items that occur naturally together, the simulation becomes indistinguishable from the experiment. If nonhandicapped students in wheelchairs are asked to perform specific tasks and then timed, the procedure would qualify as an experiment. However, if a person is asked to spend the entire day in a wheelchair and keep notes of everything that occurs, it seems more logical to call this a simulation.

Simulations are most useful in situations where observation and experimentation are not feasible or ethical. For example, a furor was caused when it was found that jury deliberations in Chicago were being taped as part of a research project. Researchers avoid controversy today by using experimental juries composed of paid volunteers. Philip Zimbardo created an artificial jail in the basement of Stanford University's psychology building using paid vol-

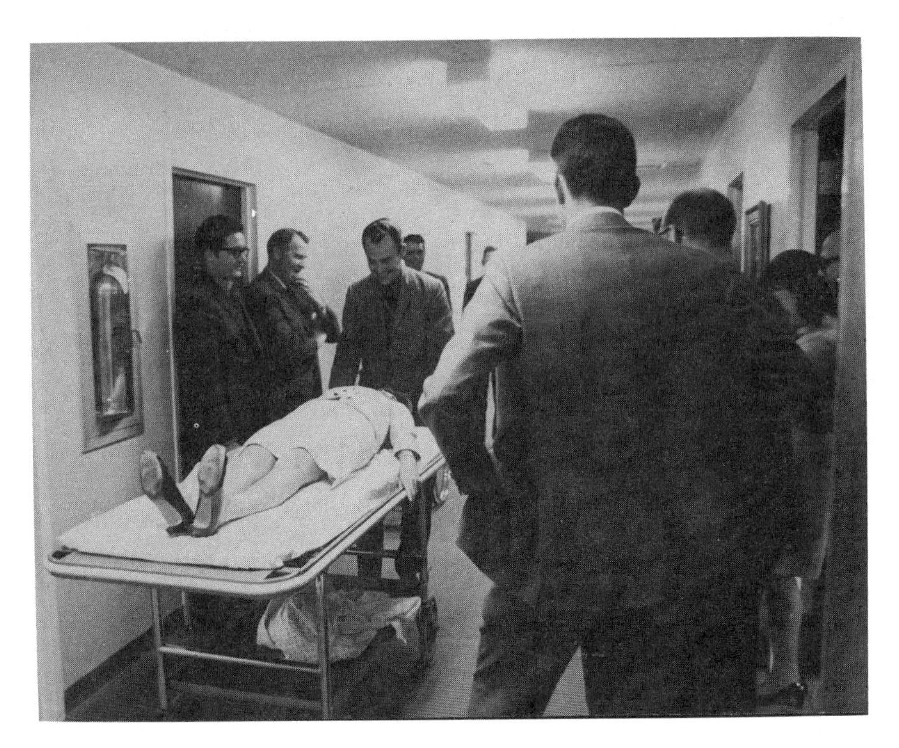

Simulation. To learn some of the problems faced by patients, hospital staff and administrators wheeled each other down the hall on gurneys and tried walking on crutches.

unteers as prisoners and guards. The experience had tremendous emotional impact upon the participants. Students playing the guards became cold, aloof, and hostile. Those who became inmates were at first angry and rebellious but in the end became sullen, bitter, and emotionally burned out (Haney, Banks, & Zimbardo, 1973).

A parallel study was undertaken by Irma Jean Orlando (1973), at a mental hospital in Illinois. Twenty-nine staff members at the hospital spent a three-day holiday weekend living as patients on a mental ward. Twenty-two of their fellow employees spent the weekend simulating the role of staff members. The admissions procedure was the beginning of a series of assaults upon the (simulated) patient's dignity. Following an intake interview, each patient was escorted to the clothing room, where ill-fitting state garments were issued, including "scuffies" (slippers associated with low status), wrong-sized underwear, and pants held up with shoestrings.

The patients were taken to the ward, where they were introduced by name to the staff, but not vice versa. There were periodic medication sessions in which the patients received placebos (sugar pills). Attempts were made to create a crowded atmosphere; the dayroom and the dormitory were small and the patients were not allowed into the dormitory during the day. The toilets were not private and there were bed checks of the patients at night. The staff further dehumanized the patients by discussing their personal problems among themselves in full hearing of the patient. Special requests by the patients were often answered, "If I let you do it, then everyone else will want to do it."

These measures proved effective in demoralizing the patients. Some were found pacing back and forth, while others sat mute in the available chairs, looking down at the floor or staring at the blank wall. Several of the patients, but none of the staff, wept uncontrollably at times. Tension on the ward kept rising, and at least one of the patients expressed fear of having an actual breakdown. Everyone filled out a brief questionnaire before the experiment ended. There was unanimity among the patients about feeling ignored, dehumanized, and rejected.

Simulation procedures have also been used in intergroup relations, where some people become minority group members for the duration of the experiment, or a class of students is randomly divided into two groups, the reds and the blues, and efforts made to see how they get along in various conflict-inducing situations. Because of the freedom allowed the participants in their role-playing, simulation procedures may be difficult for the experimenter to control, and the situation requires close monitoring to keep from getting out of hand.

Board games such as Monopoly and chess are simulated competitions. War games were used by the Prussian general staff to train military officers during the nineteenth century, and many other countries have used board games in military schools since then. Training games are also widely used in business education. There are probably as many business games as all other simulation procedures combined.

Simulation is widely used in research in architecture, city planning, and landscape design. Often it is more economical to create an artificial environment than transport people to distant locations to test their responses. Presentation modes for environmental simulations include slides, video, photographic prints, maps, floor plans, drawings, scale models, computer graphics, and various combinations of the above (Craik & Feimer, 1987).

The Environmental Design Research Laboratory at the University of California, Berkeley, uses several different simulators. A controlled environment chamber allows experimental regulation of temperature, humidity, lighting, and ventilation in a room resembling a modern office. The chamber is designed for investigation of the comfort requirements of office work. There is a separate acoustical laboratory for studying the effects of noise. For research on land use planning, the Berkeley Laboratory has an environmental simulator. This involves a remotely guided lens suspended from overhead to move through or over scale models of buildings or outdoor settings. A camera attached to the apparatus produces realistic pictures for later analy-

Simulation. Board games can be used in research on social problems. These people are playing a game designed to teach about social inequality. Some players begin the game with less money and have fewer chances for advancement than others.

sis. In an evaluation of proposed sites for an interstate highway bridge, use of this simulator permitted people living near Shreveport, Louisiana, to compare before-and-after views of the different locations and express their preferences (Atkins & Blair, 1983).

Not all environmental simulations are as elaborate or expensive as this. Other researchers employ cardboard scale models, photographs, and slides. People's placement of dolls on scale furniture produced results similar to distances real people used in conversation (De Long, 1976–1977). Landscape researchers used slides to obtain ratings of environmental quality. People might be shown slides of different parts of a coastline and asked to rate scenic attractiveness, naturalness, excitement, and so on. Methods have been developed for making composite images from several slides. This allows assessment of the same location with different environmental features.

Recent developments have brought videotape technology within the reach of many researchers. Almost any situation, from riding a bicycle to counseling, can be simulated on videotape. The technique is widely used in the human factors field, in such applications as driving simulators that now include video images of the roadway and potential accident sources for greater realism.

Limitations

Two criticisms of simulations are that they are too realistic in some instances and not realistic enough in others. The imitation jail constructed in the basement of Stanford University's psychology building was extremely stressful for the participants. Even milder forms of role-playing can be upsetting. People become so identified with their roles that they say and do things they regret later. It is difficult to predict in advance how people will respond in a role-playing exercise and thus protect them from personal revelations or actions that may be harmful to them or upsetting to others. For example, when people take part in role-playing exercises, they reveal themselves in the presence of co-workers, friends, fellow students, supervisors, and others. When the mayor and the county air pollution officer play APEX, each has the opportunity to observe the other's performance, and if the mayor is in charge later on, this may have serious implications for the air pollution officer's job. Ethical problems of simulation exercises outweigh their usefulness as a research tool in many situations. Simulation should be considered primarily when other methods are unavailable or inappropriate.

The second major criticism of simulations is that they are artificial and unrealistic. Playing games is not the same as real life. Board games such as chess and Monopoly are less a copy of actual conflict situations than another level of reality (i.e., game playing). The brief duration of play, the fixed rules, and the illusory quality of the rewards and penalties make the situation different from ordinary life.

Summary

Behavioral simulations are intended to resemble actual situations in many of their functional characteristics without being mistaken for them. Simulations include experimental games, role-playing exercises, and scale models. Compared to actual situations, simulations are economical in terms of time and money.

Limitations: When a simulation is not sufficiently realistic, it will lack validity; however, if it is too realistic, it can be stressful for the participants. Simulation is used primarily when other methods are unavailable, inappropriate, or too expensive.

References

Atkins, J. T., & Blair, W.G.E. (1983). Visual impacts of highway alternatives. *Garten und Landschaft, 8,* 632–635.
Craik, K. H., & Feimer, N. R. (1987). Environmental assessment. In D. Stokols & I. Altman (Eds.), *Handbook of Environmental Psychology* (pp. 891–918). New York: Wiley.
De Long, A. J. (1976). The use of scale models in spatial-behavioral research. *Man-Environment Systems, 6,* 179–182.
Haney, C., Banks, C., & Zimbardo, P. (1973). Interpersonal dynamics in a simulated prison. *International Journal of Criminology and Penology, 1,* 69–97.
Orlando, I. J. (1973). The mock ward. In O. Milton & R. Whaler (Eds.), *Behavior disorders* (3rd ed.) (pp. 162–170). Philadelphia: J. B. Lippincott.
UCSC Review, 1978, *3,* 3.

Further Reading

Feimer, N. R. (1984). Environmental perception. *Journal of Environmental Psychology, 4,* 61–80.
Jones, E. R., Hennessy, R. T., & Deutsch, S. (Eds.). (1985). *Human factors aspects of simulation.* Washington, DC: National Academy Press.
Wrightsman, L. S., O'Connor, J., & Baker, J. N. (Eds.) (1972). *Cooperation and competition: Readings in mixed motive games.* Monterey, CA: Brooks/Cole.

IV
Survey Design

8 Interview

Face-to-face interviews provide an excellent way of exploring complex feelings and attitudes. The following interview was designed to explore the psychological accompaniments of attitudes of truants and a matched sample of nontruants.

Question. What sorts of things do you like to do in your spare time?

Answer. I like to play football and go out with my friends. Whatever my friends like to do.

Q. Do you spend most of your time alone or with friends?

A. I hang out with a whole bunch of friends, mostly after school.

Q. Where do you and your friends usually like to go?

A. Sometimes we go to the Graduate to play video games. I don't know. . . . just find some place (seems shy and embarrassed at lack of words).

Q. What do you think about school?

A. I like it. It's better than doin' nuthin'. If it weren't for the homework, I'd like it better. I don't like my geography teacher. He's kind of senile.

For this project, there were several reasons why the interview was a better method than other techniques. If the researcher had tried to hand out questionnaires in class, many of the truants would not have been present. Second, an interview allows the researcher to pursue half-answered questions and to encourage more thorough and detailed responses. Finally, the face-to-face contact allows for observation of general appearance, overall health, personality, nonverbal behavior, and other individual characteristics.

There are times when interviews will produce more accurate information than other types of procedures that seem on the surface to be more rigorous and objective. Researchers found that interviews were more accurate in detecting alcoholism than were laboratory tests. The best of the laboratory tests used in the study detected only a third of those with serious alcohol problems while interview procedures detected 95 percent of excessive drinkers (Bernadt, Mumford, Taylor, Smith, & Murray, 1982).

Interviews constitute a rich and fascinating source of information for research. Their intrinsic interest stems from the personal interaction that is the core of the procedure. Modern technology has led to a broadening of the face-to-face concept to include interviews by telephone, video, and other extended means of communication. However, the key element of the interview is the verbal give-and-take between two persons with the questions and answers providing its form. Another way of describing an interview is that it is a "conversation with a purpose" (Bingham & Moore, 1959).

It is surprising to discover all that people are willing to talk about. With encouragement and the recognition of genuine interest on the part of the interviewer, people will reveal a great deal about themselves and about their beliefs and feelings. In collecting case histories on sexual behavior, Alfred Kinsey and his associates appealed to the willingness of people to contribute information that would be of scientific value and of help to others. The basic pitch was to altruism, but its mode varied for each group interviewed. Professional people responded when they recognized the social significance of the study. Less educated people responded positively when told simply that doctors needed their help in knowing more about these things. The researchers were also quick to show their appreciation and esteem for those who helped. The success of the studies of sexual behavior was due not only to the scholarly competence of the investigators but also to their recognition of the fundamental importance of the human interaction which occurs between the interviewer and the respondent. The Kinsey group respected their interviewers and appointed them trustees of the Institute of Sex Research and members of the policymaking board. Over a period of 22 years, nearly 300 persons were considered as possible interviewers, from whom a total of 9 was selected (Pomeroy, 1963). One outcome of this concern for the quality of interviewing is that the Kinsey studies remain an important landmark in the study of human sexual behavior.

Uses and Types of Interviews

The interview is particularly useful for the exploration of topics like truancy and sexual behavior, which are complex and emotionally loaded. Interviews are also useful in areas where opportunities for observation are limited. They can be used to assess beliefs and opinions as well as personality characteristics. A person's answers may reveal both manifest and latent content. *Manifest* content is that which is obvious and conveyed in the spoken information of the interview. It refers to what the person says. *Latent* content is the less obvious or more hidden information conveyed by hesitations and non-verbal responses, such as avoidance of eye contact, nervous gestures, or restlessness. Latent content indicates what the person means. The interviewer pays special attention when the latent content contradicts the manifest content. When interviews are recorded on videotape, the analysis of particular movements, gestures, and facial expressions, as well as changes in voice quality, hesitations ("um . . . you know . . ."), and pauses, provide a wealth of data for assessing the latent aspects of the communication. Such material may then be tabulated in a content analysis (see Chapter 11).

Interviews may focus on personality characteristics rather than on the content. One example is the *psychiatric interview,* an in-depth exploration of personality. These purposes may be combined. Social work interviews often serve the dual function of obtaining necessary factual information as well as gaining insight into a client's personality. Job interviews also combine these goals, where character as well as competence is evaluated. Newspaper and television interviews provide factual information to the viewer but also give glimpses into the respondent's personality. In watching an interview with Fidel Castro, most of us are as interested in what he is like as in what he says. Research interviews also provide this dual function. As we explore facts and opinions, we learn about the respondent.

A further advantage of the interview is that people who may be unwilling or unable to write out a long, coherent answer are often willing to say it to an interviewer. Of course this may also become a disadvantage when the interviewer encounters a long-winded respondent who wanders off the subject and won't stop talking. In such cases, the interviewer needs to be firm but polite in guiding the respondent back to the point.

Structured and Unstructured Interviews

Interviews may range from highly structured to unstructured interactions. In a *structured* or *standardized interview,* the questions are formulated beforehand and asked in a set order and in a specified manner. The structure is provided to obtain consistency from one situation to the next. In survey research where hundreds of people are contacted by several different interviewers, structured formats are essential in getting information which can be combined. The Gallup and Harris polls utilize mostly structured interviews to make their projections of public opinion.

In an *unstructured interview* the main goals are to explore all the alternatives in order to pick up information and define areas of importance that might not have been thought of ahead of time, and to allow the respondent to take the lead to a greater extent. The interviewer has a general topic in mind and may want to ask specific questions. However, there is no predetermined order or specified wording to the questions.

An unstructured interview leaves room for improvisation on the part of the researcher. Studying the use of the coca plant (the basis of cocaine) in Peru, a sensitive topic in discussions between locals and foreigners, Kirk and Miller (1985) were dissatisfied with the stereotyped replies they received. People were saying what they thought the interviewers wanted to hear. The interviewers became dissatisfied with the sameness of people's statements. As they put it, their measurement was too reliable! The discrepancy between what they were hearing and what they knew from other sources to be local practices compelled the interviewers to change their approach. To minimize the likelihood of stereotyped, consistent answers, the researchers decided to ask novel questions, such as "When do you give coca leaves to animals?" Or "How did you find out you didn't like coca?" These seemingly "silly" questions produced new and useful information.

Whether the interview will be structured (standardized) or unstructured will depend on its purpose. If information from a number of respondents is to be combined, as in an attitude survey or opinion poll, then a structured interview is desirable. An unstructured interview is desirable as a preliminary step in developing a structured form for an interview, or a written questionnaire, or may stand as part of a qualitative study.

If data are to be meaningfully combined across respondents, then the questions must be asked in the same way, thus requiring a more structured or standard format. However, some questions still may be left open-ended—for example, "What do you think about X?" and "Why do you feel that way?" The important point is that the same questions must be asked in the same manner for all respondents.

Sometimes it is necessary to design a *semi*structured interview where all respondents are asked the same questions, but the order in which they are asked differs from one person to the next. In some cases, even the manner in which they are asked varies, for example, changing the wording or sentence structure to better fit the respondent or the situation. This arrangement may be more suitable for obtaining in-depth information where the interviewer does not want to be restricted by a prescribed question order but would like the advantage of having asked the same questions of all respondents. Interviews with a cross-section of the population may require adaptation of wording and sentence structure to better fit the respondent's age or social background.

Using a semistructured interview, one loses the consistency provided by following the same procedure for all respondents. The further one moves from a structured procedure, the greater the likelihood of interviewer bias.

Telephone Interviews

Telephone interviewing has become more feasible due to various technological advances. There is random digit dialing which overcomes some sampling problems. There are statistics on telephone ownership and usage that can be used as a check on your sample of respondents (Frey, 1989). For researchers on a limited budget, there are WATS lines and special rates that have decreased the cost of long distance calls. A telephone survey of 100 households in a community is far more economical than door-to-door interviews. There are, however, problems unique to telephone interviewing, including households with unlisted numbers, answering machines, bias in terms of who is likely to be home at certain times of the day, and the possibility of intrusion during mealtimes or other personal activities.

Despite these disadvantages, there are many situations in which telephone interviewing is the most feasible and economical approach. As an example, the student organization on our campus uses the telephone to survey student opinions on current issues, based on a random sample of telephone numbers from updated student directories. With the assistance of the Registrar's Office, special lists can be obtained for different categories of students, such as those living in the residence halls or off campus, undergraduate or graduate students, and so on. Telephone contact allows excellent sampling of students plus rapid turnover time. Using five or six interviewers, the student organization can obtain interviews from a random sample of 200 students in a few days. A mail survey would take longer and be more expensive.

Depth Interviews

The *depth* or *intensive interview* is a special form of unstructured interview. The interviewer follows the respondent's answers with a request for more information at an increasing level of depth. It is this process of using the respondent's answers to delve more deeply into the topic that gives the depth interview its name.

> *Example*
> Does your school operate on a semester or a quarter system?
> How do you feel about that? (rather than "Do you like it") Why?
> Could you be more specific?

As one line of questioning runs out, the direction is shifted.

> Have you ever attended a school with a different calendar system?
> How did that work?

or

> Do you have friends at colleges with the (semester/quarter) system? What did they think about it?
> What would you do if the current system at your school were changed?

In pursuing the topic, the tone of questions remains neutral, giving the respondent as much freedom as possible to express feelings on the topic.

It requires practice and sensitivity to know when to change direction.

When a new topic emerges in the discussion, it is difficult to know whether to follow it and risk losing continuity or stay on the major theme and risk the omission of additional information. The decision should rest on your assessment of the respondent and how much shifting of questions can be tolerated. Some people are easily confused and distracted by sudden shifts. Other become bored with a direct and obvious sequence. If the respondent begins answering questions before you ask them, consider picking up the pace and expanding the scope of your interview.

How to Interview

The most effective learning of interview techniques will come from practice combined with feedback. Practice with other people before doing the actual interviews. Such role playing will give you a chance to familiarize yourself with the procedures and provide an opportunity for feedback.

Videotaping and watching a playback of yourself conducting an interview will be helpful in improving your skills. Also, watching skilled interviewers and evaluating their performance is instructive. If a group of you are planning to work on a project, practice on each other and evaluate each other's performance.

Deciding What to Ask

In many cases, the topic will not be a matter of choice but is determined instead by the demands of the situation. The interview process will be considerably simplified if you pick a topic with which most people are familiar. If your questions refer to rare experience or knowledge, you will spend a lot of time searching for people who know something about the topic. Another consideration is to select an issue that is nonthreatening. Ethical considerations require care in interviewing about illegal or taboo subjects. The respondent should be protected from any possible harmful effects of the interview, for example, against court subpoena of interview records or publication of embarrassing information. Sensitive topics must be pursued sensitively. If they are not, the likelihood of a successful completion of the interview is markedly reduced.

A final consideration is to pick a topic you find interesting. Your interest or lack of interest is likely to be conveyed to the respondent. A ho-hum reaction on the part of an interviewer quickly extinguishes any enthusiasm a respondent might have about answering the questions.

Begin with the more general and more interesting questions. By starting in a general manner, you are less likely to inadvertently influence the answers. A specific question may establish a *set* (a tendency to respond in a

Interviewing children poses special problems. This interviewer felt that getting down to the child's own level increased rapport.

particular way), that then influences subsequent answers. Remember, respondents tend to make judgments about what a questioner wants to hear. Starting with more general questions reduces the likelihood of this happening. The reason for beginning with interesting questions is fairly obvious—to engage the interest and attention of the respondent. Also, questions that might be more challenging or complicated are better asked early. The routine ones that require little thought can be placed at the end when fatigue is more likely—information covering name, birth date, place of birth, and so on. This procedure relieves some uncertainty because the respondent knows exactly what content will be associated with his or her identification.

Questions should follow one another in a logical order without abrupt changes in subject. If a change is necessary, a bridge statement from the interviewer eases the transition.

Examples
Now, to change the subject, are you in favor or opposed to ________?
Changing the subject, what do you think about ________?
The next few questions will deal with ________.

Each subsection of the interview should contain questions that move from the general to the specific.

The first questions asked should determine whether or not the respondent has knowledge of the topic you wish to cover. *Don't assume anything.* A graduate student was once interviewing a young man hospitalized for a suicide attempt. The patient explained that he had quarreled with his lover and had then become despondent. After having already asked several questions which the respondent had willingly answered, the student asked, "How long had she been living with you?" The respondent said heatedly, "It's not a she, my lover is a man." There was an awkward pause as the graduate student readjusted his frame of reference. The interview continued, but there was a loss of rapport between interviewer and respondent.

Obtaining the Interview

Before considering a household survey, you should find out if it is legal. In an effort to regulate the activities of door-to-door salespeople, many communities require a license for any sort of solicitation. Because this was designed for salespeople, it probably won't affect a legitimate research study. However, some unscrupulous salespeople use the "I'm doing a survey" approach to get a foot in the door. This has given survey research a bad name. To be on the safe side, you should check with the local police department before embarking on a community survey. The police like to know what is going on around the precinct in case they receive telephone inquiries. There are certain neighborhoods where researchers may require police protection. Under those circumstances, it would be wise to switch to another data-gathering technique or another neighborhood. Some projects send out interviewers in pairs, one of whom always waits outside while the interview is in progress.

The interviewer's task becomes easier when people expect them or at least know that the survey is being undertaken. Knocking on doors will provoke many refusals from people who are disinterested, suspicious, or afraid of strangers. An advance phone call announcing that the interviewer is coming will be useful. Another helpful device is a leaflet placed in the mailbox or under doors the day before announcing that the survey is being undertaken and explaining its goals. Having the sponsorship of a recognized community agency will make your work easier. Sometimes even government sponsorship may not be sufficient, as in this case of responses of Minnesota farmers toward interviewers from the United States Department of Agriculture:

> The farmer scowls. "I won't give you anything, I'm disgusted." The interviewer walks away empty handed. Later the farmer explains his reaction to a reporter. "When she says 'Department of Agriculture,' I see red."
>
> The interviewer is accustomed to such treatment, especially nowadays, with agriculture in a near depression and many farmers blaming Washington for much of what is wrong. Although the majority continue to discuss their business and plans with federal pollsters when asked, thousands refused, and not always politely (Birnbaum, 1982).

First impressions are important. Appropriate dress and grooming will reduce reluctance or suspicion. Make a clear and honest introduction, giving your name and institutional affiliation and an explanation of the purpose of the interview. Your manner and voice should be friendly, courteous, and nonthreatening. Overfamiliarity should be avoided. In explaining the purpose of the interview, be sure to give the person time to understand what you are saying. It may be the fortieth time you have rattled it off, but it will be the first time your potential respondent has heard it.

> *Example*
>
> Good afternoon. My name is Robin Jones and I represent the International Research Corporation. We are doing a household survey on transportation needs in the community. May I please ask you a few questions on this topic? The interview should take no more than 15 minutes of your time.

It may be difficult to obtain an unbiased sample in a household survey. Who is at home will vary by time of day. In the afternoon there will be a disproportionate number of homemakers, older people, and children. It is also difficult to interview people in a crowded household. Securing privacy is a problem, and there are the distractions of chores, children, telephones, and television.

Street Corner Interviews

To avoid the problems of a house-to-house survey, some researchers prefer to collect their interviews on street corners, in parks, shopping malls, and other public locations. Again, it is important to gain the permission of the agency or corporation that controls the space. Local police should be consulted about a street corner survey, and the management of a shopping mall regarding a survey conducted on its premises. Most of the time people are willing to oblige a worthwhile and legitimate project. However, some of them may have been burned by individuals or groups promoting political causes, religious concerns, or commercial projects in the guise of doing a survey. Frequently it will be helpful in gaining an interview if you can state factually that the management of the shopping center or the park department has agreed to let you do the survey. Such permission adds legitimacy to your study.

Stopping people walking on the sidewalk may not be as easy as it sounds. Many people are in a hurry and others have developed a tunnel vision for strangers. They simply won't see you as you approach. It is best to dress neatly and carry an obvious clipboard that conveys your purposes at a glance. If you are mistaken for a panhandler, then you are doing something wrong.

Choose a location where people can see you from a distance. That way you won't surprise them with your request. This will enable them to convey through their actions their own curiosity, willingness to help, or clear disinterest. The best sort of location is where people are seated or moving slowly. Interviewing office workers eating lunch in a courtyard outside their building will be easier than stopping them on the street as they leave the building after work. Your success rate will be higher if you pick people who are strolling or window shopping than those who are walking hurriedly toward an obvious destination. Don't try to stop people who are rushing to catch the 5:02 train or trying to get on a bus. If you can get permission from the manager, an airport where people spend hours with nothing to do is a fine place to conduct a survey. Parks, beaches, museums, and shopping areas are also good locations for interviews.

Ingenuity and a knowledge of the characteristics of the population to be studied contribute to the success of an interview. A team of architects and social scientists studying housing preferences in the Navajo nation was having difficulty obtaining information through conventional means. The Navajo had been surveyed so often with so few results that they were uninterested in answering questions. Knowing the Navajo interest in traveling about their territory, the research team gained much of its information from hitchhikers. They picked up and interviewed riders as they drove them to their destinations. Under these circumstances, people were very cooperative (Snyder, Stea, & Sadalla, 1976). Sometimes you have to be creative in approaching the people you want to interview (see Boxes 8-1 and 8-2).

Setting the Stage

A major consideration in putting another person at ease is to be at ease yourself. Role playing and practice interviews will increase your confidence. Respondents generally are more likely to give candid responses when they are convinced that their responses will be kept confidential and when they perceive that no moral evaluation of them is being made by the interviewer. This observation was borne out by the researchers in the Kinsey studies of human sexual behavior. Despite the extremely intimate and sensitive nature of their questions, they obtained remarkable cooperation. The need for a nonjudgmental attitude and for assurances of confidentially cannot be overemphasized. Human beings are very good at detecting deceit. If the inter-

BOX 8-1. Creative Technique for Interviewing Bicyclists

The easiest way to find bicyclists, the research team decided, was to seek them out on the streets. A door-to-door survey would be unlikely to turn up more than a handful of dedicated riders. The first problem was to get the cyclists to stop long enough to be interviewed. The research team stationed themselves on a clear stretch of road frequently traveled by bicyclists. They placed a sign 100 feet ahead of them, "Bicycle survey ahead." This aroused the curiosity of most cyclists. The researchers carried obvious clipboards and had big signs across their chests that read "Bicycle survey". Most cyclists, alerted by the earlier notice and spotting the two sign-bedecked interviewers frantically waving their arms, pulled over, and answered the questions. The novelty of the approach proved successful in that situation. Undoubtedly, if the procedure were repeated many times, the novelty would wear off. How would you go about conducting a survey about the attitudes of motorcyclists? Would you use the same procedure as in the bicycle survey or would you have to do something different?

viewer is making strong moral judgments, it will come across during the interview. Such feelings must be reduced prior to the interview. This may be accomplished by recognizing the legitimacy of other people's views. Another means of reducing the emotional intensity of feelings is through discussion, which may make a topic less sensitive. If your own feelings remain strong, perhaps another person should be found to do the interviews.

BOX 8-2. Interviews in a Multimethod Study

Psychologist Donald Norman (1983) combined several research techniques in his studies of people's use of hand calculators. Norman gave people a series of arithmetic problems that were to be solved on a hand calculator. Participants were requested to "think aloud" as they solved the problems while Norman watched and recorded their words and actions. When all the problems were completed, Norman questioned the participants about the methods used and about their understanding of the calculator. Although all the people were reasonably experienced with the machines on which they were tested, they also seemed to have a distrust of the calculator or in their understanding of its operation. As a result, they would take extra steps or decline to take advantage of calculator features that were known to them. Those who had experience with several different calculators often confused features of the different machines. They were unsure which feature applied to which calculator and had various superstitions about the operations of the keyboard. For example, people would hit the CLEAR button several times saying such things as "You never know—sometimes it doesn't register." A similar pattern applied to the use of the ENTER button. People would push it repeatedly, often while commenting that while they knew this to be excessive, "It doesn't hurt to hit it extra" or "I always hit it twice when I have to enter a new phase—it's just a superstition, but it makes me feel more comfortable." The combination of observation and interview used in this procedure was superior to either method used alone.

Find a place for the interview that is free from distraction and serves to maintain confidentiality. Your respondent is not likely to be candid within earshot of others. Also, telephone calls and other interruptions interfere with the respondent's train of thought. Be sure your own needs for a writing surface and a comfortable location for recording are taken care of before you start. Otherwise, you may have to interrupt the session and you may miss important information.

Nonverbal Aspects

A degree of eye contact is important in establishing rapport between interviewer and respondent. However, don't stare. An unbroken or penetrating gaze will make the respondent uncomfortable. On the other hand, the absence of eye contact makes you appear untrustworthy or uncomfortable. Again, the best preparation for interviews is practice and getting honest feedback from classmates, colleagues, or friends.

The rest of your face, as well as your eyes, communicates your feelings, particularly those of approval and disapproval. Respondents are very likely to pick up even flickers of feeling. Thus, it is very important for an interviewer to show an attentive but nonjudgmental interest. As most of us are not trained actors, we will probably perform best when asking questions that interest us but that do not arouse intense emotions. For emotionally sensitive questions, a considerable amount of practice is essential in desensitizing feelings.

While conducting practice interviews, notice the set of your mouth, your body posture, and your position in relation to the respondent. When do you lean forward? When do you lean back? What might these behaviors communicate? Also, pay attention to your hands. Do you tend to cover you mouth when asking a question? The latent message of that movement is that you don't really want to ask it. Just as videotapes are widely used in training psychotherapists, so are they of value in providing feedback for research interviews. Observe the professional interviewers on television. Notice the effects of their behavior upon the respondents, and vice versa.

The quality of the interview data rests on the interaction between interviewer and respondent. Good questions can be ruined by improper techniques. Poor questions may be salvaged by a good interviewer. Remember, the purpose of an interview is to obtain information about another person's beliefs, opinions, feelings, or attitudes. Responding to the person's answers in a positive way is essential. One need not be vigorous in approval, but rather, through an occasional nod or eye contact, indicate attention and respect for the views expressed.

Interviewing people of different ethnic background, nationality, or social class brings special problems of nonverbal communication. In some cultures, for example, avoidance of eye contact is a sign of respect rather than an indication of untrustworthiness. Physical distance between interviewer and respondent has different meanings for persons of different backgrounds. Northern Europeans and Anglo-Americans often become uncomfortable within the close range quite acceptable for those of Latin background (Hall, 1959). Familiarity with the slang expressions of another social stratum may be necessary. A respondent will soon lose interest if the interviewer can't understand what is being said; for example, probing the drug culture requires knowledge of the various terms used for the drugs and persons associated with them.

Characteristics of race and sex as well as social class can affect the interview. However, the heart of the issue is the degree to which these characteristics will detract from obtaining candid responses. Persons of the same social or ethnic background also introduce a bias of conformity to expected attitudes. The quality of the interview depends more on the interviewer's capacity to convince the respondent of a nonjudgmental attitude and a sincere interest in and respect for the respondent's point of view than upon the specific characteristics of race, sex, or class. However, in some situations, these characteristics may be particularly influential and require a matching or deliberate contrasting of interviewer and respondent characteristics.

Pacing and Timing

A major mistake of beginning interviewers is the failure to allow enough time for the respondent to answer. The interviewer, especially after having formulated the questions, becomes very familiar with the outline and may forget that it is totally new to the respondent. A person needs time to think about each question and to prepare an answer. The interviewer must learn to be comfortable with long silences, yet recognize the point at which the silence is no longer productive and is making the respondent uncomfortable. If an answer is not forthcoming, a reformulation of the question may be appropriate. Sometimes the discomfort caused by silence may be used to the interviewer's advantage. It may lead the respondent to be more self-revealing than he or she originally intended.

To practice how to tolerate delay, look at a watch or clock with a second hand and observe the passage of 30 seconds. As silence, it may seem a conspicuously long period of time. Yet, while attempting to formulate an answer to a complicated or controversial question, 30 seconds is hardly time at all. The main point is for you to become less self-conscious about silence.

The interview should not be allowed to drag. When the respondent has finished answering a question, move on to the next. Otherwise, boredom, irritability at the interviewer's unpreparedness, or rambling into irrelevant comments may ensue. The interview should move at a pace that is rapid enough to retain interest but slow enough to allow adequate coverage of the topic.

Probes

When a respondent gives an unclear or incomplete answer, it is necessary to probe for additional information. A *probe* is a question or comment designed to keep the person talking or to obtain clarification. If the respondent does not seem to understand the question or strays from the topic, it may be sufficient to repeat the question. If that does not work, or if only a partial response is given, other probes are necessary. An expectant pause accompanied by the usual facial expressions of waiting, or repeating the respondent's reply followed by a pause, should draw added comment. Failing that, use neutral questions or comments or simply ask for further clarification.

> *Examples of probes*
> (Interviewer repeats all or part of question)
> Anything else?
> Could you tell me more of your ideas on that?
> What do you mean?

It is very important that probing questions or comments not bias the response. They should not direct a person to a particular answer.

Recording the Information

The degree to which notes reflect exactly what was said depends on how the interview will be used. When the information is the final source of data, a precise transcript is essential. The most common means of recording is to write down the respondent's replies exactly as spoken. This generally requires a rapid notation system in order to avoid the respondent's losing interest. On the other hand, it is flattering to have someone write down one's statement in detail. Thus, there is a positive aspect even though some delays and interruptions are required.

The type of interview and scoring system will help determine how much the interviewer should record. Slips of the tongue, pauses, and defensive gestures, that would be very important in a clinical interview probably won't be worth recording if the responses are going to be put into categories of agree/disagree or satisfied/dissatisfied. The task of classifying answers into fixed categories is called *coding*. This task becomes more time consuming when the coder must wade through a lot of irrelevant detail. *Verbatim responses* are those written down exactly as spoken and would include ungrammatical statements, requests for more information, and so on. *Paraphrased responses* are the interviewer's impressions of what the respondent meant. Things that are irrelevant, unnecessary, or ungrammatical are omitted.

Example

Verbatim statement: "I think . . . umm . . . that he is . . . well . . . doing the best he can. They are good . . . they are trying hard. It is not easy to be in the White House for any man."

Interviewer's paraphrase: The President is doing a good job.

Such interpretations by an inexperienced interviewer would be risky. It is probably better for the beginner to write down more, rather than less, detail. What is irrelevant can be ignored at the time of coding. When dealing with groups of people who have a special way of expressing things, such as adolescents or professional engineers, it is important to record statements exactly as they are said. On the street, "bad man" may be a term of respect. Direct quotations can be used to make a report more readable. Instead of talking about the gang members' attitudes toward authorities, the researcher can express their attitudes in the words of the gang members.

Various shorthand techniques, long practiced by stenographers, may be used. Kinsey and his associates developed a very elaborate code system that required 3 months to learn and 6 months to practice with ease. No written words were used, only mathematical signs and numbers. These coded data have never been transcribed into a longhand or typewritten account. The coding during the actual interview served several functions. It enabled the interviewer to record all pertinent data without slowing up the interviewing and risking loss of rapport. It preserved the confidence of the record since the code has never been publicly explained. It facilitated the transfer of the data from the original interview notes to statistical analysis. If a written record had been used, then the data would have required coding anyway. Thus an intervening step was omitted by coding the information as it came from the respondent.

Tape recorders provide an advantage of precision and also will record hesitation phenomena, the various "ahems" and "ahs" and "ers" made by the respondent. However, transcription is a long process, taking four or five times the amount of time spent in the interview itself. Many respondents are made uncomfortable by the presence of a tape recorder, and confidentiality may be compromised. Thus, tape recorders are not recommended for most interviews.

Where replies are likely to be lengthy, various shorthand codes may be developed and used. In many interview situations, a self-developed notation is adequate, especially if notes are gone over immediately following the interview, with scribbles and shorthand terms correctly spelled out while the conversation is fresh in mind. A common convention is the use of a hash mark (/) to indicate where a probe has been made.

Example

Question: What do you think about forced retirement at age 65?

Answer: Well, it is probably a good idea in general, but then for some people not so good. / Well because some people are still pretty sharp and have a lot of experience to offer. / Like judges and politicians, things like that.

Other conventional marks may be adopted in an interview project. The University of Michigan Survey Research Center (1976) uses a standardized set of abbreviations for probes (see Table 8-1).

In many instances an exact word-for-word transcription is not necessary, nor is it necessary to code the material for statistical analysis. In conducting in-depth interviews as a preliminary step to questionnaire construction, discovering categories of response is more important than recording every single word or anticipating statistical analysis. Jotting down notes from significant responses will be sufficient.

Tact and Diplomacy

You must learn when to probe and when to end a line of investigation. This is particularly important in the less structured situations. Depending on format or procedural constraints, an interviewer may choose to vary the order or phrasing of the questions, adapting them to the situation and mood of the respondent. Another aspect requiring tact is to recognize when a respondent is being untruthful or wandering off the subject. A direct indication of mistrust on the interviewer's part may terminate the interview. On the other hand, false information is of little value unless its assessment is a purpose of the interview. There are no simple guidelines to deal with these problems. If you are working in an area where misinformation or irrelevancies are likely to crop up, then you should plan beforehand how you will deal with them.

If the respondent becomes tense or hostile, it may be desirable to terminate the session, either permanently or until another time. People occasionally take out their resentment about an issue on the interviewer. Because of

TABLE 8-1. Abbreviations for Probes Used by the University of Michigan Survey Research Center

Interview's probe	*Standard abbreviation*
Repeat question	(RQ)
Anything else?	(AE or Else?)
Any other reason?	(AO?)
Any others	(Other?)
How do you mean?	(How mean?)
Could you tell me more about your thinking on that?	(Tell more)
Would you tell me what you have in mind?	(What in mind?)
What do you mean?	(What mean?)
Why do you feel that way?	(Why?)
Which would be closer to the way you feel?	(Which closer?)

the number of violent confrontations that occur during questioning, police in several cities have switched to bulletproof clipboards. You will probably not need to consider such extreme measures, but you should be alert to the possibility and, over time, the inevitability of someone becoming upset and angry. When this happens, soothe the respondent's feelings, and end the interview politely and quickly.

Ending the Interview

At the outset, tell the respondent how long the interview will last and stick as close to that limit as possible. If it is a paid interview, the means and time of payment should be explained. On completion, express appreciation for the person's time and effort. The respondent should be encouraged to ask any questions about the research, and should be told how the results will be used, and where and when they will be available.

> *Example of a closing sequence of comments*
> Thank you very much for your time and effort. Is there anything you would like to add to the interview? (pause) The results of our survey will be sent to the State Department of Transportation. You will receive a check for $5.00 in the mail in about 2 weeks. If there is any problem, please telephone. You have my card. Do you have any further questions about the survey? (pause) Thank you again. Goodbye.

The ethics of research require as much feedback as possible to the respondents about the nature, purposes, and intent of the project. Common courtesy requires acknowledgment of the value of a person's time and effort.

Analyzing the Results

Keep your survey data in a safe place. The unspoken fear of researchers is graphically depicted in a newspaper report of an apartment house fire:

> Alan Rosin, who had lived in one of the buildings for 14 years, had just completed a political poll. Now, the damp and partially burned surveys containing 1,352 interviews lay in a mound of debris on the sidewalk. "All the (data) were in my apartment. Now they are all over the street," he lamented. (Lambert, 1984).

When the interview is the primary source of research data, it becomes necessary to devise systematic ways of summarizing information. The first step is generally to provide for transcription and quantification. Transcription refers to putting the responses into clear form for data analysis, such as transcribing answers from a tape recorder to typed copy, or recopying handwritten notes clarifying abbreviated terms. Quantification means presenting the results in numerical form.

Coding

Coding is the process by which categories of responses are established for open-ended questions. It is the means by which lengthy statements are reduced and sorted into specific response categories. The easiest items to code are sex, age, religion, occupation, and other categorical characteristics. These are categorical in that a person clearly belongs in one or another category (e.g., male, below age 25, and Catholic). Code categories are generally numerical. However, the choice of numbers is arbitrary.

> *Example*
> Question: When in your life did you decide upon your present career?
> Coding categories:
> 1 = Childhood (10 yrs. or under)
> 2 = Adolescence (11 through 17 yrs.)
> 3 = Young adulthood (18 through 24 yrs.)
> 4 = Later years (25 yrs. or older)
> 5 = Still undecided about career

It is very important to keep a record (in duplicate) of the coding numbers used. Your data summary won't make sense if you have 354 cases of code 6 under religion and cannot remember whether these are Catholics or Protestants.

The coding process is more complicated for more complex answers. Box 8-3 illustrates the coding process for an item taken from college students' recollections of their experiences of puberty. Another example of coding can be found in Chapter 11.

Reporting Interview Results

You are not bound to the interview format when presenting the results. It is not necessary to give the answers to question 1 first, then question 2, and so on. Instead, the order should reflect the importance of the findings. The most clear and significant results should be described first. Areas of less agreement and importance come next. Trivial or irrelevant findings need not be mentioned.

BOX 8-3. Coding Open-ended Responses

A. The first step in analyzing the responses is to read them over and generate a set of categories into which they can be classified. The following categories were constructed from the answers to the question: "What social changes or events did you, at the time, associate with puberty?"

Code #

1. Interest in other sex; social activity with other sex—dances, dating, parties, going steady.
2. Increased independence and/or responsibility.
3. Self-consciousness; attention to appearance.
4. Need for social acceptance, seeking friendship, joining clubs and organizations.
5. Withdrawal from family, problems in communication with family.
6. Increased social sensitivity (i.e., awareness of cliques, feeling social pressure, conformity concerns).
7. Other (if too many of the responses fall into this category, it should be made more specific, as those above).
8. No changes, or can't recall.

B. The next step is to check the reliability of those categories. Find a person willing to assist you and provide a description of the coding categories. Independently score a sufficient number of interviews to test the reliability of the scoring system. If the rate of agreement is low, then revise your coding categories and check reliability again.

C. In this example each response is scored using more than one category. For example, the social changes mentioned might have been both interest in the other sex (code 1) and withdrawal from family (code 5). In other cases you may wish to assign responses to one category only. The choice will depend on the nature of the answers, and whether it is feasible to place them in mutually exclusive categories.

Code #	*Male respondents*
6,4	More aggressive, needed to prove self and to be accepted.
1,5	Increased social activity, less interaction with parents and family.
3,1	Shy, withdrawn. Girl friends—or at least I tried!!!
	Female respondents
1,6	Dances, relationships with guys more than girls, pressure to conform, and I began a relationship with a guy when I was not really ready.
8	Not much.
5,4	Trouble communicating with family, turned more to friends.

D. Following the coding, responses are summarized by category:

Code #	*Coding category*	*Number of females mentioning category* (n = 14)*	*Number of males mentioning category* (n = 14)
1.	Interest in other sex	4	5
4.	Social acceptance	2	3
3.	Self-consciousness	3	1
6.	Social sensitivity	3	1
2.	Independence/responsibility	1	1
5.	Withdrawal from family	1	1
7.	Other	0	1
8.	No changes, can't recall	2	1

*n refers to the number of respondents in the sample.

Indicate the number of people contacted, the number of refusals, replies that had to be discarded for one reason or another, and the final number of respondents. Discuss the means by which people were contacted, the dates, time of day, and something about the training and background of the interviewers.

Remember to safeguard the identity of your respondents. This can be done through the use of code names rather than real names and the removal of statements that contain identifying information not essential to the reader's understanding of the study.

Limitations

What people say is not always what they do. The information obtained in interviews is limited to the spoken content and to inferences made by the interviewer. The data are highly subject to bias introduced by the human interaction of the interview process. Interviewers may unintentionally encourage or discourage the expression of particular facts and opinions. While no research method is absolutely free of subjectivity, the interview is more open to bias than most other research methods. However, this is not to say that bias is inevitable; rather, to warn that great care in constructing the question format and in training interviewers is essential if valid information is to be gained. The necessary care may be costly. The need for training, coupled with the time-consuming aspects of the interview itself, creates an economic disadvantage. A written questionnaire distributed to a hundred people takes a small fraction of the time required for individual interviews. The coding of the open-ended questions used in interviews is time-consuming and also expensive. Results from more than one interviewer may not be combined unless each has proceeded in the same manner.

Summary

The key element of an interview is the verbal give-and-take between interviewer and respondents. Interviews are particularly useful for exploring complex and emotionally loaded topics. Interviews can be used to explore beliefs, opinions, and personality characteristics. In a structured interview, questions are formulated beforehand and asked in a set order in a specified manner. In an unstructured interview, the interviewer has in mind a general topic, but not set questions or a predetermined order to the questions. In an in-depth interview, the interviewer follows the respondent's answers with a request for more information at an increasing level of depth. The classification of the respondent's answers into categories is called coding. Interview techniques are learned best through practice combined with feedback.

Limitation: Compared to questionnaires, the interview is time consuming and expensive. The responses are subject to bias introduced by the human interaction during the interview process.

References

Bernadt, M. W., Mumford, J., Taylor, C., Smith, B. & Murray, R. M. (1982, February 6). Comparison of questionnaire and laboratory tests in the detection of excessive drinking and alcoholism *The Lancet,* 8267, 325–326.

Bingham, W. V. D., & Moore, B. V. (1959). *How to interview* (4th ed.). New York: Harper.

Birnbaum, J. H. (1982, December, 22). Farmers need savvy, as do agents who ask for data. *The Wall Street Journal,* p. 1.

Frey, J. H. (1989). *Survey research by telephone* (2nd ed.). Beverly Hills, CA: Sage.

Hall, E. T. (1959). *The silent language,* Garden City, N.Y.: Doubleday.

Kirk, J., & Miller, M. L. (1985). *Reliability and validity in qualitative research.* Beverly Hills, CA: Sage.

Lambert, M. (1984, October, 8). Apartment fire leaves only sorrow. *Sacramento Bee,* p. B1.

Norman, D. A. (1983). Some observations on mental models. In D. Gentner, & A. L. Stevens (Eds.), *Mental models* (pp. 7–14). Hillsdale, NJ: Erlbaum.

Pomeroy, W. B. (1963). Human sexual behavior. In N. L. Farberow (Ed.), *Taboo topics.* New York: Atherton Press.

Synder, P. Z., Stea, D., & Sadalla, E. K. (1976). Socio-cultural modifications and user needs in Navajo housing. *Journal of Architectural Research,* 5, 4–9.

University of Michigan, Survey Research Center (1976). *Interviewer's manual* (rev. ed.). Institute for Social Research, University of Michigan.

Further Reading

Douglas, J. D. (1985). *Creative interviewing.* Newbury Park, CA: Sage.

Fowler, F. J., & Mangion, T. W. (1989). *Standardized survey interviewing.* Beverly Hills, CA: Sage.

Ives, E. D. (1980). *The tape recorded interview: A manual for field workers in folklore and oral history* (rev. ed.) Knoxville, TN: University of Tennessee Press.

Lavrakas, P. J. (1988). *Telephone survey methods.* Newbury Park, CA: Sage Publications.

Stano, M. E., & Reinsch, N. L., Jr. (1982). *Communication in interviews.* Englewood Cliffs, NJ: Prentice-Hall.

Weinberg, E. (1971). *Community surveys with local talent.* Chicago: National Opinion Research.

9 The Questionnaire

Sir Francis Galton, a nineteenth-century versatile English genius, was fond of measuring things. He set up a booth at the 1884 International Health Exhibition to measure people's physical responses, including their reaction time, strength of pulling and squeezing, hearing, seeing, and color perception. The testing was later moved to the South Kensington Museum in London. The apparatus stood on a long table at one end of a long, narrow room. After paying a three-penny admission fee (which helped to support the research), each individual performed the various tests while an attendant recorded the responses. The goal was to determine the range of human sensory capabilities. Galton hoped to test the entire population of Great Britain so that the nation would know the exact state of its mental resources. He was never able to fully realize his dream. By the time the program ended, however, he had tested 9337 individuals, quite an achievement for his time.

Galton was also interested in visual imagery—the ability to picture things in one's mind. This was something that could not be observed or tested directly. To assess visual imagery, Galton presented people with a list of written questions about their ability to picture breakfast that morning. Was the image dim or clear? Was it in color or black and white? Were the colors distinct and natural? This was one of the first uses of the questionnaire in psychological research. Respondents were asked to rate the vividness of their images along a scale from 0 (no image at all) to 100 (clear and distinct as the original).

Galton found that artists reported better imagery than scientists. Some scientists, in fact, doubted that imagery existed. They felt that the expression "Seeing in the mind's eye" was only a figure of speech and that the very notion of mental pictures was fanciful. Galton discovered not only important differences in the way people think but also the value of the questionnaire as an instrument for studying behavior that could not be observed or experimented on directly (Galton, 1907).

A *questionnaire* is a series of written questions on a topic about which the respondent's opinions are sought. It is a frequently used tool in *survey research*—the systematic gathering of information about people's beliefs, attitudes, values, and behavior. There are two general types of questionnaires: self-administered, which respondents fill out themselves, and interviewer administered, in which the interviewer asks questions and records the responses. The self-administered form is more efficient in time and effort. For example, copies of a questionnaire can be distributed to 100 employees, filled out by them, and collected within an hour's time. Conducting 100 individual interviews would be much more arduous and time consuming. However, a self-administered questionnaire requires clear instructions and a very careful wording of items. The most difficult aspects of a questionnaire are its construction and the interpretation of the results. Distributing and scoring a well-constructed questionnaire are usually easy (although the amount of data to be processed can be extensive). Overall, it is difficult to surpass a questionnaire for economy. That is why it is such a popular and widely used research tool.

Questionnaire Construction

There are two general aspects to every questionnaire: its content and its format. The *content* of a questionnaire refers to the subject matter. The *format* pertains to its structure and appearance—how the items are worded, their appearance on the page, and the form used for answering the questions.

Content

In general, it is best to restrict a questionnaire to a single issue. If you want to find out what college students think about the cafeteria, stay with that topic. Don't ask about teaching methods or the adequacy of the library. If others are interested in these topics, let them do another survey. It will be easy for students sitting in the cafeteria to see the relevance of questions about the food and service.

Is yours the first survey on this issue? Sometimes a beginning researcher is halfway through a survey before discovering that someone else studied the same topic a year earlier. The cafeteria manager or a long-term employee may be a good source of information on this. It is surprising how many surveys are finished, filed, and forgotten. You can save considerable time and effort if the previous survey developed suitable questions. You can always add new questions, but try to include some of the old ones. Wherever possible, retain the same wording. If the previous questionnaire asked whether the employees were "helpful," don't change the word to "courteous" even though you believe this means the same thing. Even a slight change in wording can invalidate the comparison between answers from two surveys.

Assuming you are going to construct your own questionnaire, you will start by using two other methods: casual observation and interview. The purpose of casual observation is to learn the *range of activities* about which questions must be asked. You need to know what goes on in order to ask questions about it. This requires brief inspection rather than systematic observation with detailed categories. Interviews should accompany the observations to learn the range of opinions students hold regarding the food. At this stage, a loose, open-ended interview is preferable. This will avoid suggesting answers to people. Replies will be used primarily to find out which topics should be included in the questionnaire. The following brief list of questions would probably be appropriate for a casual interview on this topic:

> Hello, my name is _________. The Campus Planning Committee is doing a survey of student attitudes about the cafeteria. I would like to ask you a few questions. This will only take a few minutes. Your answers will help us improve food service on campus.
>
> 1. What do you think of the food here?
> 2. What do you like most about it?
> 3. What do you like least about it?
> 4. What do you think about the service and facilities?
> 5. Is there anything else you would like to say about the food or the service?

These general questions would not be sufficient for a formal interview. However, the purpose is not to use the answers as primary data but only to learn what questions should be asked on the questionnaire.

Format

Open-ended and Closed Questions

There are two major categories of questions: open ended and closed. With *open-ended questions,* the respondents write in their own answers:

> *Example*
> What do you like most about this cafeteria?
> What do you like least about this cafeteria?

Closed questions, also known as *multiple-choice questions,* ask the respondent to choose among alternatives provided by the researcher.

> *Example*
> What do you think of the salads here?
> (like) (dislike) (indifferent)
> What do you think of the cost of the meals here?
> (too expensive) (very reasonable) (about right)

An open-ended format is desirable (1) when the researcher does not know all the possible answers to a question, (2) when the range of possible answers is so large that the question would become unwieldy in multiple-choice format, (3) when the researcher wants to avoid suggesting answers to the respondent, and (4) when the researcher wants answers in the respondent's own words. Multiple-choice (closed) answers are desirable (1) when there is a large number of respondents and questions, (2) when the answers are to be scored by machine, and (3) when responses from several groups are to be compared.

The beginning researcher often prefers open-ended questions because they allow respondents more freedom to answer as they please. This infatuation with open-ended questions tends to be short lived. Once the researcher attempts to compare one person's answers with another's, and the replies from one group with those from another group, the disadvantages of open-ended questions become quickly apparent. People tend to ramble in their answers to open-ended questions. It is therefore necessary for the researcher to go through lengthy statements in order to pick out the respondent's intent or meaning. Often it shows more respect for people's opinions to let them classify their answers themselves as positive, negative, or neutral than for the researcher to do this for them.

Salience refers to the importance of an issue in people's minds. Open-ended questions are very useful for determining the salience of opinions. It is generally assumed that in answering an open-ended question, those items that stand out in a person's mind will be mentioned first. Another method for determining salience is to ask respondents to *rank* a list of items in terms of importance. This can be done using either an open-ended or a closed format.

> *Example (open ended)*
> List those qualifications you would look for in hiring a secretary. Place a 1 next to the most important qualification, a 2 next to the second most important, and so on, until all qualifications on your list are ranked.

> *Example (closed)*
> Rank each of the following qualifications in terms of its importance for a secretary in your office (1 = most important).
> ____ Typing speed
> ____ Typing accuracy
> ____ General clerical skills
> ____ Computer skills

____ Taking dictation
____ Telephone answering skills
____ Meeting the public
____ Other (please specify)

Ranking versus Rating

Rank-order questions tend to confuse many respondents. They are also more difficult to analyze. Some people will list only their top choices or use check marks instead of numbers. Others interpret the scales in the wrong direction, using 5 for the most important skill and 1 for the least important skill, even though the instructions say the opposite. Other people will list two number 3s and omit number 4 entirely. The larger the number of items to be ranked and the more complicated the format, the more mistakes will be made. With more than 10 items, rating is more efficient than ranking. Sometimes a combination of rating and ranking can be used.

Example

Rate each of the following qualifications in terms of its importance for a secretary in your office. Please use the following scale:

1. Very important
2. Moderately important
3. Slightly important
4. Not at all important

____ typing speed
____ typing accuracy
etc.
____ meeting the public

Now place a check next to the single most important qualification and two checks next to the second most important qualification.

Matrix Questions

Arranging the items and answers in a matrix is very efficient. By matrix, we mean using answer headings across the top, and listing the items down the side, as illustrated in this example:

For each statement below, please indicate whether you Strongly Agree (SA), Agree (A), are Undecided (U), Disagree (D), or Strongly Disagree (SD).

	SA	*A*	*U*	*D*	*SD*
The food in the cafeteria is well prepared	( )	( )	( )	( )	( )
Lines are too long	( )	( )	( )	( )	( )
Table arrangements encourage conversation	( )	( )	( )	( )	( )

The degree of detail desired by the researcher determines the number of response categories. If earlier interviews revealed that the students either strongly liked or strongly disliked the cafeteria food, there would not be much point in using 10 categories. Because the students do not expect the cafeteria to be the equal of a fine restaurant, there is little reason to include an "excellent" category. Three terms would probably be sufficient: "good," "ok," and "poor." Those terms will take up little space, will be easily understood by the respondents, and will serve to distinguish between positive, neutral, and negative opinions. In other situations where a finer differentiation of opinion is required, 5 or 7 categories can be used.

Examples
Excellent, good, satisfactory, poor, terrible
Strongly agree, moderately agree, slightly agree, undecided, slightly disagree, moderately disagree, strongly disagree

Measuring the Middle Position

The middle position refers to a neutral or undecided stance. When a respondent checks "undecided," it isn't always clear what this means. Sometimes it means a balance between positive and negative feelings on the issue; other times it means a lack of interest or knowledge on the topic. Some investigators eliminate the middle position, creating a forced-choice item where the respondent must agree or disagree, or take some kind of a stand. Other investigators prefer to use "don't know" as a middle-ground response. Having a middle category will decrease the number expressing a clear opinion at one end or the other. Not surprisingly, middle category responses occur more often when people don't have strong feelings on the issues (Schuman & Presser, 1981). Although it is possible that people who actually have an opinion might use the middle ground, others may, in fact, feel neutral about an issue. We favor providing the middle alternative in order to avoid forcing a false appearance of opinion one way or the other, and out of respect for the respondent's right to neutral feelings. Several students interviewed in the cafeteria expressed these types of opinions.

The food's okay, I guess.
What do you expect of cafeteria food?
It's better than what you get at Jack-in-the-Box.

A neutral category seems appropriate for vague opinions like these. Some terms useful for the middle category are: "neutral," "undecided," "neither good nor bad," "satisfactory," "OK." Often it is helpful to include the additional response category "not applicable (NA)" for questions that are not relevant. Students who never eat a particular food item may feel more comfortable checking NA than saying that it is neither good nor bad.

Example
What do you think of the desserts served here?

Good OK Poor NA

Combining Question Types

Following the chapter on multimethod approach, it will come as no surprise that the authors recommend a combination of closed and open-ended items, ratings, and ranks, rather then relying on a single type of item. The customary procedure is to use open-ended, general questions at the beginning of a questionnaire, followed by more specific, multiple-choice rated and ranked items. Coupling several types of items provides checks on each. In a survey designed to learn student attitudes toward library reading areas, students were first asked an open-ended question, "What do you like most about the library reading areas?" The most common response was the quiet in the library compared to the student's dormitory room or apartment. Later, students were asked to rate specific aspects of the library environment as "excellent," "satisfactory," or "needs improvement." Paradoxically, the item of library environment that provoked the largest number of complaints was the noise.

This illustrates the value of including both open-ended and closed questions in the same survey. If only the general question about why the student came to the library had been asked, the researchers probably would have concluded (incorrectly) that the noise situation was good. Yet the ratings indicated just the opposite. Students came to the library in search of quiet, but serious noise problems still existed.

Number of Questions

Beginning researchers often include too many items in a questionnaire. In their desire to omit nothing, they forget that the respondents will become fatigued and lose interest as they plod through an unending barrage of questions. For most purposes, the shorter the instrument, the better. Avoid complex formats, such as, "If yes, skip to question 9, part B. If no, answer question 7, part A."

Wording Your Questions

Your questions should be clear and meaningful to the respondents. Terms should not be too difficult. Don't overestimate the vocabulary level of your respondents. Define all difficult or jargon terms. It is helpful to include synonyms. Instead of asking a question about *nuclear power,* you can ask about *nuclear or atomic power.* The synonym may reach some people who miss the primary term.

Avoid loaded terms, scare words, and phrases that immediately raise a red flag. Don't ask questions about *dangerous* drugs, *noisy* trucks, or *excess* government spending. Such adjectives produce a reflexlike negative response. If you want to know what people think about cocaine, ask them directly; don't ask about dangerous drugs. *Noise* is a technical term for the unpleasant aspects of sound. Asking employees to evaluate the *sound level* in the office is preferable to asking them about noise, which by definition is unpleasant.

Avoid double-barreled questions. Ask about one thing at a time. Don't ask about the cost *and* efficiency of municipal services. It is preferable to divide this into separate questions, one concerned with efficiency and the other with cost.

Whenever possible, avoid phrasing questions in the negative. It is surprising how often people confuse "do" and "don't," "will" and "won't," "can" and "cannot." If you must use a negative, be sure to underline it or have it printed in italics so that it will not be missed.

Example

Of all the classes that you took in high school, which was the single class that you liked *least?*

Question wording must be culturally sensitive. When a questionnaire on American high school students was adapted for use in Australia, several questions had to be reworded. In Australia, the term *college* refers to what would be the last 2 years of high school in North America. Questions about "going to college" were changed to "attending a tertiary institution." As background information, Australian students were asked if they were of Aboriginal descent, a question that would have little meaning in the United States. Also, the Australian students were asked if they had ever lived on a "farm or station," the latter term referring to what would probably be a ranch in the United States. Changing the wording makes it difficult to compare the results of different surveys, but may be necessary in order to make the questions understandable to a new group of respondents.

A common mistake in writing questions is to use jargon terms. Many employees do not know the meaning of "FICA" or "SSI" even though social security payments are deducted from their paychecks every week. Some words have more than one meaning. When a survey team asked residents of a fundamentalist community about the need for government to regulate profits, they were surprised to find exceptionally strong resistance to the idea. Subsequent interviews disclosed that some people thought the question referred to regulation of prophets!

Balance

Balance refers to the neutrality of questions, or providing sufficient items so that the number leaning toward one view are balanced by an equal number leaning toward the other view. The goal of balance is to counteract implicit influence from the questions themselves. The following shows imbalance in questions about cafeteria food.

Is the meat too salty?
Are the vegetables overcooked?
Is the coffee watery?

This set of questions is unbalanced at two levels. Individually, the items imply a direction to the answer. Balance can be provided by asking "Is the coffee too weak, too strong, or about right?" The second level of bias is that the implied responses for all the items are in the same direction—that the food isn't very good. An alternative to balancing each individual item is to balance them overall. For example, leave the implicit bias in the question "Are the vegetables overcooked?" but balance it with another item of the reverse implied bias, such as "Is the meat properly prepared?" Providing balance not only reduces the effects of a possible response bias, it also makes the survey itself seem more fair, and thus more difficult for partisans of one view or another to ignore or dismiss.

Don't be discouraged if your questionnaire isn't perfect on the first try. It is extraordinarily difficult to anticipate all sources of ambiguity and confusion. Experienced survey researchers always construct a preliminary version, pretest it, and then revise it before distributing the actual survey. The secret of learning to write clear questions is practice, feedback, more practice, more feedback, and still more practice. No one writes a perfect questionnaire the first time.

Checklist for Evaluating Questionnaire Items

1. Is the question necessary? How useful will the answers be?
2. Is the item clear and unambiguous?
3. Will the respondent be competent to answer the question as asked?
4. Will the respondent be willing to answer the question as asked?
5. Have double-barreled questions been eliminated or revised?
6. Is the item as short as possible, while remaining clear and precise?
7. Do the multiple-choice questions provide a comprehensive set of choices? Do they include a "don't know" or "not applicable" category? Is there an "other" category, if appropriate?
8. Is the answer likely to be affected by social desirability (saying the "right thing")? If so, can the question be altered to reduce this bias?
9. Have negatives such as "no" and "not" been eliminated insofar as possible?
10. Are the questions balanced so that the number of favorable items equals the number of unfavorable items?

Layout

A self-administered questionnaire must begin with an introductory statement, present the questions in an easily read and easily answered format, and close with a note of thanks or appreciation. For interviewer-administered questionnaires, instructions at the beginning will help to ensure consistency from one interview to the next.

Introductory Statement

At the top of the questionnaire, briefly describe the purpose, identify the person or group conducting the survey, request assistance, and provide general instructions. For example,

> The purpose of this questionnaire is to learn about attitudes toward the cafeteria. The survey is being conducted by Ms. Jones and Mr. Smith, social science students at ——— College. We would very much appreciate your assistance in answering the questions below. Please do not write your name on this form in order that the replies remain anonymous.

Additional instructions might be included. For some rating scales, the respondent is asked not to dwell on the choice, to just check the first response that comes to mind. In other cases a well-considered response is requested.

Question Order

Begin with factual, noncontroversial questions. These help to establish a good relationship that can smooth the way for more difficult or controversial questions later.

General questions on a topic should precede specific questions, as in this example.

1. What do you think of this playground?
2. Is there enough play equipment?
3. Do you feel that any of the play equipment is dangerous?

This sequence avoids suggesting danger to the respondent on the first two questions. If danger is mentioned in the first two answers, it is likely to be a salient issue and influence subsequent answers.

Maintain a logical order understandable to the respondent. Routine items such as age and gender can be included at the end. The questionnaire should *not* begin with overly personal or sensitive questions regarding illicit activities, sex, or controversial religious or political opinions. If such items are to be included, begin with the less controversial ones.

Answer Format

There are various answer formats in which a respondent inserts a checkmark, number, or letter. The response blanks that are placed either to the right or left of the question usually take one of five forms: ______, /‾/, ( ), □, or ○.

With regard to neatness and appearance, the □ has been rated best, and the /‾/ was the worst, and the left-hand side was best for presenting the response blanks (Major, Jacoby, & Sheluga, 1976). However, for typed questionnaires, the ( ) or ______ are easy and acceptable. Another acceptable practice is to have the respondents circle their choice.

Arrange the response blanks to permit easy tabulation. This is especially important if the answers are to be entered into a computer. Put the answers where the operator can find them. Don't scatter them around the page. Insert computer code numbers close to the answers.

Example

What is your class in school?	Fr Soph Jr Sr 1 2 3 4	(1)
Are you a full-time or part-time student?	Full-time Part-time 1 2	(2)
What is your sex?	Male Female 1 2	(3)
What is your age?	___ years	(4–5)

If the questionnaire is precoded for computer scoring, make these codes as inconspicuous as possible and provide an explanation to the respondent, for example, "Please ignore the numbers in parentheses, which are for computer use only."

Closing Statement

At the end of a self-administered questionnaire ask for any other comments or suggestions, and then thank the respondents for their time and effort—a simple THANK YOU VERY MUCH may be sufficient.

Pretesting

The first draft of your questionnaire will need revision. The need for changes should be accepted gracefully. No matter how carefully you phrase the original questions, there will still be some words that are difficult or unclear, some topics left out. The impressive economy of the questionaire is partially offset by the researcher's inability to clarify the meaning of terms. Elaboration is easy in an interview, almost impossible on a questionnaire. Because there may be no person physically present to help explain the meaning of difficult terms or confusing questions, you must anticipate all possible sources of error. An adaptation of Murphy's Law* applies here:—The slightest opportunity for respondents to go wrong means that some of them will.

The best way to reduce ambiguity is to pretest the questions. Try them out on a group of people who are asked the items *and,* in addition, asked to

*Murphy's Law: If something can go wrong, it will.

comment on their wording and clarity. The first draft of the cafeteria questionnaire might be given to eight customers and two cafeteria employees. The respondents would be told that this is a trial version of the survey instrument and that we would like their opinions on the food *and* the questions. Pretesting can be done in a short period of time. There is no need for detailed sampling or statistics. A small sample should be sufficient unless serious problems are found, in which case a major revision and further pretesting will be necessary. Pretesting can be done orally. Try it out on a few people who will not be in the final sample but are similar to that sample.

A basic rule in questionnaire construction is that the first draft should never be mimeographed or printed. Write it down in longhand or type it, make 10 copies, but no more. Once you have a large stack of printed questionnaires in front of you, it is easy to overlook minor errors and inconsistencies and to continue to use the copies because they are available. If you do this, you will regret it afterward. It is easier to change a first draft than to interpret people's answers to ambiguous questions.

Reproducing the Questionnaire

Eye appeal is important. Make sure all questions are legible. Don't squeeze things together. Use a second sheet of paper if necessary for a clear presentation. A nicely designed and printed questionnaire will increase reader interest. With the advent of low-cost photocopies, this is your most likely means of questionnaire reproduction, although mimeograph or ditto, if available, are cheaper. Before you distribute your questionnaires, go through them to see that pages are in the correct order and there are no blank pages. Numbering the pages will make it easier for you (and the respondent) to notice when pages are out of order or one is missing.

The readability of two-sided copies depends on the thickness of the paper, so try it and see whether the print shows through before committing yourself to multiple copies. If you use the back, be sure to write PLEASE COMPLETE THE REVERSE SIDE at the bottom of the first page. Even with that warning, you can assume that in a self-administered questionnaire, some respondents will fail to complete the back page. The same expectation applies to multiple pages. If the questionnaires are being picked up individually, it is wise to make a general scan to see that all pages have been filled out. (See Box 9-1.)

In deciding on the number of copies needed, consider your sample size and request additional copies. You will need these not only for follow-up or spoilage, but also for data tabulation, discussion copies, as appendices for reports, samples for other researchers, and so on. Later reproduction of a few copies can be expensive and needlessly time consuming.

BOX 9-1. Steps in Constructing a Questionnaire

Content

1. Exploratory interviews of an in-depth type and/or casual observation, if appropriate.
2. Decide on aspects of problem to be covered. Be specific.
3. Generate items.

Format

4. Decide item wording, for example, open-ended vs. closed, etc.
5. Reexamine items (see Checklist for Evaluating Questionnaire Items, p. 136.)
6. Have a knowledgeable colleague or instructor look at the questions.
7. *Pretest*—pilot test the questionnaire *before* putting it into final form.
   a. Check for confusing or ambiguous items.
   b. Estimate the time needed for administration.
8. Revise questionnaire in accord with pilot results. If new items are added, be sure to pretest them.
9. Final editing—include clear instructions to the respondent or the interviewer.

Distributing the Forms

If the questionnaire is properly constructed and brief, you will be pleasantly surprised at how quickly it can be distributed, filled out, and collected. This is the stage at which the economy of this instrument is most evident. It may take more than a month to get permission to conduct a survey among office workers, another few weeks to write and pretest the questions, get them approved by management, and have the forms printed, but only a few hours to distribute them, have them filled out, and collected. Once you have the instrument already available, a second survey can be carried out in a wondrously short time.

Since a questionnaire is to be answered in writing, it is more easily administered to people sitting down than to those standing or on the move. One would probably choose an interview rather than a written questionnaire for customers in a shopping center or teenagers at the Dairy Queen. The questionnaire is ideally suited for office workers, students, library readers, and others who are seated at desks or tables.

There are various methods of distributing questionnaires to potential respondents. Questionnaires can be circulated at a group meeting, handed to people individually, or mailed. Each approach is likely to work better in some situations than in others. *Sampling* is an important consideration in distributing a survey. This topic is discussed in Chapter 16.

Group Meetings

When an organization, club, or class meets regularly, you have a good opportunity to distribute a questionnaire. If the survey is brief and relevant to the group's purpose, it may be possible to have the questionnaires filled out during the meeting and returned directly to you. There is likely to be a problem of a biased sample if only a small number of members attend because they may not be typical of the total membership.

Individual Distribution

Individual distribution is efficient when the behavior studied involves solitary people rather than groups or organizations. Researchers studying wilderness use have employed this approach in combination with mailing. The questionnaire is given to people as they enter or leave the area, perhaps in a parking lot or at the trailhead, with the request that the questionnaire be filled out and placed in a designated collection box or returned through the mail in an attached postage-paid return envelope. When this approach was used among people engaged in river recreation in Vermont, the return rate for completed surveys was 61 percent, which is very good for a mail survey (Manning & Ciali, 1980). The people contacted had a direct connection with the topic of the survey that contributed to the good return rate. We would not recommend handing questionnaires to people walking by on the street or placing them on automobile windshields with a request that they be filled out and returned by mail. Most of these questionnaires are likely to end up in the street or in a trash can. A positive personal relationship with the researcher or sponsoring agency is crucial for a good return rate.

Mail Surveys

A mail survey is efficient for covering a large geographic area quickly. It is expensive to send researchers to all parts of a city or state to conduct surveys. The advantages of a mail survey are its low labor and travel costs and complete standardization. There is no need to train interviewers, and all respondents receive the same questions posed in the same manner. A mail survey also provides more anonymity to the respondents than is possible in a personal interview. A follow-up study of former employees may secure more valid information when conducted by mail without requiring names to be associated with the responses. If contacted directly, individuals might fear that their opinions would be used against them. However, there will be many other situations where controversial opinions can be explored more fully through a personal interview than in a mailed questionnaire.

Disadvantages of a mail survey are its impersonality, low return rate, slowness, and financial cost. Many commercial firms and political organizations send out fake surveys as promotional tools. Respondents are asked to answer questions deliberately intended to create favorable attitudes toward a product or organization. This misuse of survey methodology has led many householders to classify all mail surveys as junk mail.

A mail survey of households randomly selected from a city telephone directory will probably result in a 10–33 percent return rate, thus requiring the researcher to send out from three to five times as many questionnaires as are needed for the final sample. There is also a problem of potential response bias when only a small percentage of questionnaires is returned. The return rate is likely to be much higher if the recipients have a prior interest in the topic of the survey. For example, a mail survey of former high school alumni accompanied by a cover letter from the school principal is likely to yield a return rate in excess of 60 percent of delivered questionnaires.

A mail survey is slower than a telephone survey, which can be done in a few days. Questionnaires take several days to be delivered by mail and weeks before the bulk of replies return. A mail survey is not appropriate when timely information is needed. The costs of printing, mailing, and return postage may seriously deplete the budget of a small organization or agency. Box 9-2 shows an example of a mail survey.

BOX 9-2. Recycling Survey (Johnson, 1984)

A questionnaire was mailed to a sample of 350 households randomly selected from a city phone directory along with a stamped return envelope and a cover letter from the sponsor of the project. Of the 350 sent, 20 were returned by the post office as undeliverable and 140 were returned completed, yielding a return rate of 42 percent on delivered questionnaires. Here are some of the items.

Recycling Survey

To be filled out by the person most likely to make recycling decisions on behalf of the household. Please check the appropriate box or fill in the blank.

Section I. This section asks questions about your current recycling activity.

1. How often does your household currently recycle the following items?

	Do not recycle	Weekly	Every two weeks	Monthly	Other (specify):
Newsprint	☐	☐	☐	☐	☐ ________
Aluminum	☐	☐	☐	☐	☐ ________
Glass	☐	☐	☐	☐	☐ ________
Tin	☐	☐	☐	☐	☐ ________
Other ________	☐	☐	☐	☐	☐ ________

---

2. For those materials you currently recycle, approximately what percentage of the total amounts used in your home are recycled?

	Do not recycle	1–20%	21–40%	41–60%	61–80%	81–100%
Newsprint	☐	☐	☐	☐	☐	☐
Aluminum	☐	☐	☐	☐	☐	☐
Glass	☐	☐	☐	☐	☐	☐
Tin	☐	☐	☐	☐	☐	☐
Other ________	☐	☐	☐	☐	☐	☐

5. If the weekly curbside program required you to sort glass by color (clear, green, and brown) instead of mixing the colors, what is the likelihood of your recycling glass?

Definitely would *not* participate	Probably would *not* participate	Undecided	Probably would participate	Definitely would participate
☐	☐	☐	☐	☐

Only apartment dwellers answer question #8

8. If containers for recyclables were placed near the refuse containers in your complex, what would the likelihood be of your recycling the following materials?

	Definitely would *not* participate	Probably would *not* participate	Undecided	Probably would participate	Definitely would participate
Newsprint	☐	☐	☐	☐	☐
Aluminum	☐	☐	☐	☐	☐
Glass	☐	☐	☐	☐	☐
Tin	☐	☐	☐	☐	☐

Section III. This section asks background questions that will help us in interpreting this survey. Your honesty is greatly appreciated and all answers are confidential. [Demographic questions follow on original questionnaire.]

Mail surveys necessarily require a literate population, an efficient mail service, and a receptive climate for filling out and returning questionnaires. In the mid-1980s, housing researchers in the Soviet Union did not feel that they could use a mail survey. At the time, there was no tradition for people to receive questionnaires in the mail, fill them out honestly, and return them. Also, mail service could be unreliable. Despite the need for a large sample—4000 families in 20 Soviet cities—the team used individual interviews rather than a mail survey, which would have involved less time, effort, and expense (Niit, 1989).

Increasing Response Rates

Notifying people about the survey beforehand will increase cooperation. For example, placing a sign in the library telling people that a survey is being undertaken and explaining its purposes will arouse interest and encourage cooperation. Office workers could be sent a note describing the goals of the survey and why it is important for them to participate. No matter how fine your motives, people are doing you a favor when they invest time and effort answering your questions. Therefore, try to convince them of the usefulness of their answers. This can be done *both* in the advance letter and in the introductory paragraph of the questionnaire.

Helpful Hints for a Mail Survey

Obtain current addresses. Use the most recent version of the telephone directory or other information source.

Send questionnaires first-class mail so that the post office will return the undelivered questionnaires. First-class letters are more likely to be noticed than bulk mail.

Hand address the letters and use attractive stamps. Some market researchers believe that a stamp placed at a slight angle on the envelope increases the return rate.

Arrange for a newspaper article describing the survey to appear shortly before the questionnaires are sent out. The article can describe the nature of

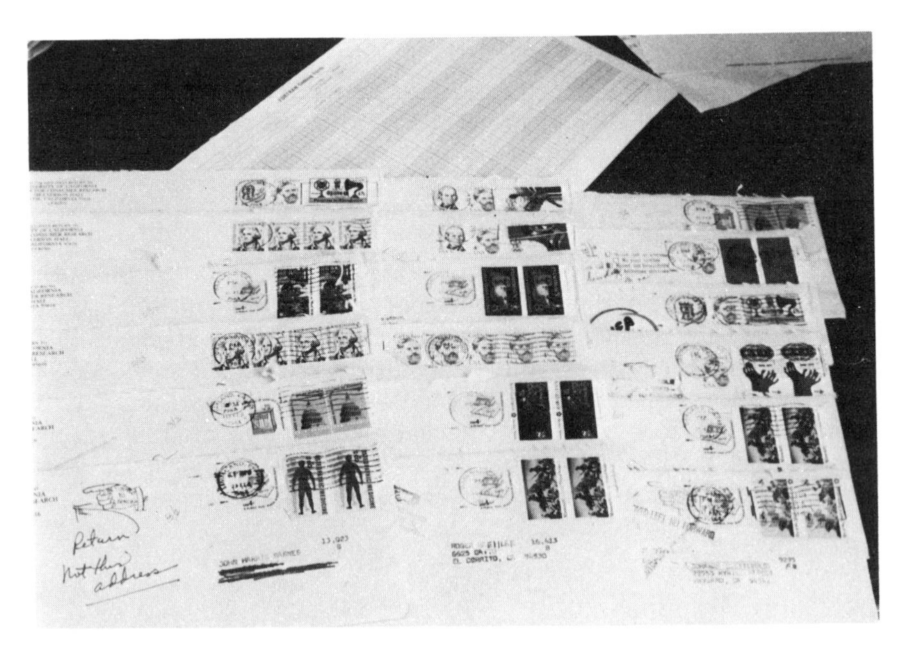

Undeliverable surveys will be returned if mailed first class. Note the use of attractive stamps to increase respondent interest.

the survey, sponsorship, what information is being collected, how it will be used, and sample selection (see Box 9-3).

Postcards can be sent before and after mailing to increase return rate. Several days beforehand, you can send a postcard describing the purposes of the survey. A week after sending the questionnaire, a follow-up reminder can be mailed asking people to please return the surveys, and providing an address to which people can write if they mislaid the original survey and would like another copy.

Example

Dear Resident,

About a week ago you were sent a questionnaire about your attitude toward various types of recycling programs. If you have not yet sent in your questionnaire, please do so as soon as you can so that our data will be truly representative of your area. If you have already mailed in your questionnaire, please accept this as a thank you for your cooperation.

Sincerely,

(signature)
(title and affiliation or sponsor)

Scoring

Scoring a properly constructed questionnaire is quite simple. If the instrument is brief and there are fewer than 100 respondents, it may be easier to score by hand than by computer. Any open-ended questions must be coded

---

BOX 9-3. Newspaper Announcement of a Survey (Birnbaum, 1982)

Major Farm Survey Launched

The U.S. Department of Agriculture is launching a major survey this month on farm production expenses, debts and assets, crop and livestock sales and costs of production.

Letters are going out to about 24,000 farmers representing a broad cross-section of American agriculture. Interviewers will then visit the farms between January 21st and March 8th to conduct the farm costs and returns survey.

Survey results will be used to determine the economic well-being of farmers.

"This is one of the best ways for farmers to get out their story," said economist Jim Johnson of USDA's Economic Research Service.

Responses to each survey are confidential.

The results will be reported to Congress, farmers, and the public.

[Similar information can be provided to radio and TV stations prior to the start of the survey. Advance publicity should increase interest and levels of cooperation.]

Addressing envelopes by hand increases the likelihood of a response in mail surveys.

by hand. See Chapter 8 for a detailed account of coding. An additional benefit of hand scoring is that you stay closer to the data and get a better sense of the responses to the questions. However, if the number of items and respondents is large, or if you wish to make many comparisons among the questions, the computer may be desirable. Nothing surpasses a computer for speed and accuracy in finding out how the opinions of clerical employees who have been employed for 10 years and who live between 5 and 10 miles from the office compare with those of clerical employees who have been employed less than 5 years and live 10–20 miles from the office. If you expect to use a computer in analyzing your results, these comparisons should be planned from the start and the questionnaire designed for computer analysis.

Interpretation

Up to now, everything has gone well. You've written your questionnaire, pretested it, revised it, distributed, and collected the final version. You are now the proud owner of 96 completed questionnaires. Two people didn't finish the questionnaire and another didn't follow the instructions. One person didn't take the questions seriously and wrote funny answers. This last is the moron factor, and the response should be disregarded. As long as it is

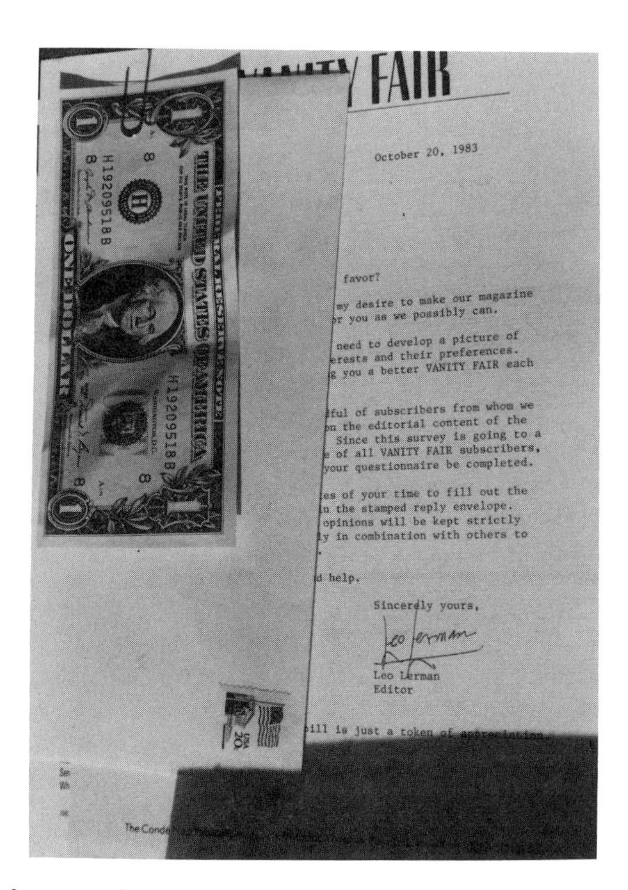

FAIR

October 20, 1983

favor?

my desire to make our magazine
r you as we possibly can.

need to develop a picture of
rests and their preferences.
g you a better VANITY FAIR each

ful of subscribers from whom we
n the editorial content of the
Since this survey is going to a
of all VANITY FAIR subscribers,
your questionnaire be completed.

es of your time to fill out the
n the stamped reply envelope.
opinions will be kept strictly
y in combination with others to

d help.

Sincerely yours,

Leo Lerman
Editor

ill is just a token

This survey from a national magazine was accompanied by a crisp new one-dollar bill as an inducement, along with a stamped return envelope.

not done to bias the results, it makes sense to discard bizarre answers. A person who puts an exclamation point after "sex" or writes "I am for it" is probably kidding. If some of the person's other replies look humorous, it is probably best to discard the entire questionnaire. On the other hand, someone putting a question mark after "age" may be shy about revealing how old he or she is. Some of your respondents will *always* omit one or two items. Either the question is not clear or the item was missed in going through the questionnaire. Don't be concerned if you don't have replies to all the questions. When you convert to percentages, there will be no problem.

When you deal with people who are rushed or busy, you can expect that some of them won't finish the questionnaire. Generally the replies from the completed portions can be added to the total sample without significant bias. This is not true of people who object to your questions. Even a mild question can upset some people. For example, an employee may use the questionnaire to vent hostility toward an employer. Before discarding these replies, it is best to check with a colleague or co-worker to see that you are not biasing your results. Occasionally an apparently humorous response is meaningful and a refusal is appropriate. Even though you have promised your respondents anonymity, some of them won't believe you. This is not totally unrealistic if they don't know you. Abnormal conditions can provoke abnormal responses. Do not expect to get honest answers from people who believe that the survey results will be used against them.

Removing the two incomplete questionnaires, the person who didn't follow the instructions, and the joker, our sample in the cafeteria survey now consists of 96 respondents—a fairly good ratio. If we had wanted to finish with a perfect 100, we could have aimed for 115 completed questionnaires and put 15 aside as replacements for any rejections. For our purposes, however, there is no significant difference between 96 and 100 completed replies. Assuming we have hand-scored the questionnaires—the quickest and easiest approach—we will probably be presenting our results in overall percentages, as in Table 9-1.

This is a straightforward presentation of the data. For many purposes, it is all that is needed. However, if you want to know how the attitudes of male freshmen living off campus compare with female freshmen living on campus, it may be awkward to sort the questionnaires for each analysis. The computer is a very efficient way of making these detailed comparisons, but unless you are careful, the computer can overload you with data. It will cost only a few extra pennies to ask the computer how the replies of 17-year-old Protestant men compare with those of 19-year-old Catholic women. We advise against asking trivial questions. The computer does not indicate which results are important and which are trivial.

TABLE 9-1. Student Ratings of the Cafeteria

	Percentage rating		
Aspect of the Cafeteria	*Good*	*OK*	*Poor*
Meats	10	55	35
Vegetables	7	44	49
Salads	53	38	8
Desserts	43	35	22
.			
.			
.			
Seating arrangement	22	62	17
Sound level	0	32	68

Describing Survey Results

People's likes can be as significant as their dislikes. The point of a survey is not to reveal only failure and frustration. Success and satisfaction should also be identified. A beginning researcher often identifies important results with surprises. Can a survey be worth anything if the results are not shocking? The answer is clearly yes. A survey that finds there is no problem may be just as important as one that identifies a source of tension. For example, if your survey tells you that students are satisfied with the cafeteria food, this is a perfectly valid and useful finding. You have found that your cafeteria is better than most. There still may be some room for improvement in other areas, such as noise and crowding.

There is nothing more frustrating to the reader of a report than to have every percentage described. You need not go through all the questionnaire items, one by one. Don't bury your reader under a mountain of statistics. Instead, select the most important results and present them first. Some findings may be unimportant or too ambiguous to have meaning. Don't exaggerate the importance of small differences. Often the similarities among various percentages are more impressive than small differences. The body of the report should be devoted to the major findings. It is realistic to assume that most readers will look only at the summary. It is still necessary to make the actual data available for those readers who want them. The best place to include detailed percentages is in the appendix, where they will not interrupt the flow of the report. Here is the summary from our cafeteria survey:

Report on Students' Attitudes Toward the Cafeteria: Summary

To learn students' attitudes toward the food and service, a 25-item questionnaire was constructed, pretested, revised, and distributed to 100 students in the cafeteria during lunch on Thursday and Friday, June 7–8, 1990. Three-quarters of the respondents had purchased cafeteria food and one-quarter brought their meals with them. The main reason students brown-bagged their lunches was the cost of the food. Areas of most satisfaction with the cafeteria were size of portions (very ample), the green salads (tasty and nutritious), cleanliness, fast service, and the possibility of going back into the line for seconds. The problems indicated by at least one-third of the sample included quality of the meats (poor), tastiness of the vegetables (overcooked), noise in the table area, and the drab decor of the cafeteria. On most of the other items, people were satisfied but not particularly enthusiastic.

This summary presents the highlights of the report. It includes what people liked as well as what they disliked. Specific complaints are placed in the context of the overall general level of satisfaction. This is accurate as well as diplomatic. It would have been biased and shortsighted to list only the complaints and omit the compliments. A good summary, like a good report, describes the full range of attitudes, including weak and noncommittal responses. The summary should be brief and to the point. The discussion comes in the body of the report, and the detailed percentage totals appear in the appendix. The raw data, the 100 questionnaires, remain in your files until you run out of room.

Comparison of Interviews and Self-Administered Questionnaires

In deciding whether to do interviewing or to distribute self-administered questionnaires, carefully consider your project needs and resources in time and money. Self-administered questionnaires are generally more economical in that they require less time and assistance in scoring, provided that they have been designed with a closed-ended response format. Administration is much quicker. They also have the advantage of offering anonymity, so they may be more desirable for investigating sensitive attitudes or behaviors.

Interviews have the advantage of allowing for observation in addition to answers to questions. Interviewers can assess nonverbal behaviors as well as elements of the setting. Even telephone interviews allow for recording intonation and comments. The interview is less likely to be incomplete and yields a higher return rate. It is better for dealing with complicated issues.

Limitations

A questionnaire is of little use with respondents who are very young, very old, infirm, or uninterested in the topic. As a written document, it is not appropriate for people on the move or who are busy with other activities. The typical questionnaire with multiple-choice answers evokes "bare bones" responses. You can learn the general structure of the situation but not the details. Follow-up interviews will help to make sense out of the trends.

A poorly worded question can create the appearance of attitudes where none exist. People may believe they are helping the researcher by providing answers that he or she wants to hear. Questionnaire replies are more useful for identifying attitudes than for predicting behavior. Do not confuse people's opinions as expressed on a questionnaire with their behavior.

Questionnaires strike many respondents as impersonal, mechanical, and demeaning and the response categories as limited, artificial, and constraining. Unless the researcher asks precisely the right questions, the information will not be very useful. Although this chapter has emphasized the economy of the questionnaire, the work involved in writing questions, pretesting the instrument, and tabulating the responses should not be underestimated. If opinions from only a small number of respondents are needed, the open-ended interview, in which a few general questions are followed by specific questions tailored to the respondents' replies, will be more economical.

Summary

A questionnaire is a series of written questions on a topic about which the respondents' written opinions are sought. Because it is so economical to administer and score, the questionnaire is widely used in behavioral research.

Open-ended questions provide space for the respondents to write in their own answers. With closed or multiple-choice questions, respondents choose from among alternatives provided by the researcher. Open-ended questions are desirable when the researcher does not know all the possible answers to a question, when the range of possible answers is so large that the question would become unwieldy in a multiple-choice format, when the researcher wants to avoid suggesting answers to the respondent, and when the researcher wants answers in the respondent's own words. Multiple-choice answers are desirable when there is a large number of respondents and questions, when answers are to be scored by computer, and when answers from different groups of individuals are to be compared.

Salience refers to the importance of an issue in people's minds. Open-ended questions are well suited to determining the salience of an issue. In a multiple-choice format, salience can be determined by asking the respondent to rank or rate a list of answers in terms of importance.

In many cases you will need a middle-response category such as "neutral" or "undecided," and need to include a "nonapplicable" or "don't know" response.

Check all items for clarity. Avoid double-barreled and negative questions. Provide balance. Always pretest the questionnaire before putting it into final form.

The questionnaire is of little use with respondents who are very young, very old, infirm, or uninterested in the topic. The typical questionnaire with multiple-choice answers evokes "bare bones" responses. Questionnaires are more successful in identifying attitudes than in predicting behavior.

References

Birnbaum, J. H. (1982, December 22). Farmers need savvy, as do agents who ask them for data. *The Wall Street Journal,* p. 1.

Galton, F. (1907). *Inquiries into human faculty and its development.* London: Dent.

Johnson, S. L. (1984). *Source separation recycling programs.* Unpublished master's thesis, University of California, Davis, CA.

Major, B. N., Jacoby, J., & Sheluga, D. A. (1976). Questionnaire research on questionnaire construction. *Purdue Papers in Consumer Psychology,* No. 166.

Manning, R. E., & Ciali, C. P. (1980). Recreation density and user satisfaction: A further exploration of the satisfaction model. *Journal of Leisure Research, 12*(4), 329–345.

Niit, T. (1989, August). Family lifestyles in new housing developments. Paper presented at symposium, *The meaning and use of home and neighbourhood,* Alvkarleby, Sweden.

Schuman, H., & Presser, S. (1981). *Questions and answers in attitude surveys: Experiments on question form, wording, and context.* New York: Academic Press.

Further Reading

Converse, J. M., & Presser, S. (1986). *Survey questions: Handcrafting the standardized questionaire.* Beverly Hills, CA: Sage.
Frey, J. H. (1989). *Survey research by telephone* (2nd ed.). Beverly Hills, CA: Sage.
Labaw, P. (1986). *Advanced questionnaire design.* Cambridge, MA: Ballinger Publishing.
Rossi, P. H., Wright, J. D., & Anderson, A. B. (1983). *Handbook of survey research.* New York: Academic Press.
Schuman, H., & Presser, S. (1981). *Questions and answers in attitude surveys: Experiments on question form, wording, and context.* New York: Academic Press.
Sudman, S., & Bradburn, N. M. (1982). *Asking questions.* San Francisco: Jossey-Bass.

10 Attitude and Rating Scales

What Is a Scale?

Originally from the Latin word *scala,* meaning a ladder or flight of steps, a scale represents a series of ordered steps at fixed intervals used as a standard of measurement. Rating scales are used to rank people's judgments of objects, events, or other people from low to high or from poor to good. Commonly used scales in behavioral research include attitude scales designed to measure people's opinions on social issues, employee rating scales to measure job-related performance, scales for determining socioeconomic status used in sociological research, product rating scales used in consumer research, and sensory evaluation scales to judge the quality of food, air, and other phenomena. These scales provide numerical scores that can be used to compare individuals and groups.

There are various methods for making ratings. With *graphic rating scales,* the respondent places a mark along a continuous line. The ends and perhaps the midpoint of the line are named, but not the intervening points. The person can make a mark at any point along the line. The score is computed by measuring the distance of the check mark from the left end of the scale.

Example

Place a checkmark somewhere along the scale to indicate the quality of this loudspeaker system.

√

Excellent (Score 7.5 cm) Terrible

A *step scale* requires the rater to select one of a graded series of levels. Intermediate points cannot be used. The intervals can be letters, numbers, or adjectives. For scoring purposes, these may be numbered from 1 to 5.

Example
How would you rate the quality of this loudspeaker system? (circle one)
excellent very good good poor terrible

Comparative rating scales are commonly used on recommendation forms. The person is asked to compare the applicant with others in the same category.

Example
Compared to other students in the class, how does this student rate in terms of motivation for graduate work?
Top 1%___ Top 5%___ Top 10%___ Top 25%___ Top 50%___ Bottom 50%___

Levels of Measurement

Quantification, the process of assigning numbers to events, ratings, or behavior, occurs at a particular *level* of measurement. These levels can be nominal, ordinal, interval, or ratio. *Nominal* measures are qualitative or categorical, providing no information about quantity. Examples of nominal measures are male and female, or homeowner, renter, and other. Numbers may be used, but only to represent categories (e.g., 1 = male, 2 = female). An *ordinal* scale provides additional information about size or direction. Street addresses are ordinal measures. They indicate direction, but provide no certain information about the distance between individual buildings. A ranking of contest winners into first, second, and third place is ordinal if there is no description of the size of the differences between them.

An *interval* scale possesses the qualities of an ordinal scale, plus the additional characteristic of equal intervals between scale points. Such scales contain units similar to a temperature scale, on which the difference between 85 and 87° F is comparable in degrees to the difference between 47 and 49°. Interval scales for measuring attitudes can be constructed using the Thurstone procedure described later in this chapter. An example of an interval attitude scale is shown in Table 10-1.

Ratio scales not only have equal intervals, but have the additional property of an absolute zero point. This permits mutiplicative comparisons. Grade point average (GPA) is a ratio measure (a student who received all Fs would have a GPA of 0). Time, distance, and physical qualities such as weight, age, and size, are easily expressed in ratio measures.

Most subjective rating scales are likely to be ordinal rather than interval or ratio. In an opinion survey, we can say that someone who strongly disagrees is more opposed than someone who merely disagrees, but we don't know how much difference there is between the two attitudes. The level of measurement in a study influences comparisons and generalizations that are justified, as well as the selection of statistical tests. Here is a summary of the characteristics of the four levels of measurement:

Scale type	Distinguishing features	Example
Nominal	Characteristics assigned to categories. No underlying continuous dimension.	Hair color: blond vs. brunette
Ordinal	Characteristics can be ordered along an underlying dimension, but no information is provided about the distance between points.	Rank ordering preferences—1st, 2nd, and 3rd choice
Interval	Comparable intervals can be determined along the underlying dimension.	Thurstone-scale; subjective judgments of equal distance.
Ratio	Has characteristics of an interval scale, plus an absolute zero point.	Measures of income or GPA.

Attitude Scales

An attitude scale is a special type of questionnaire designed to produce scores indicating the overall degree of favorability of a person's attitude on a topic. It is constructed so that all its questions concern a single issue. Attitude scales are used most often in attitude change experiments. One group of people is asked to fill out the scale twice, once before some event, such as seeing a propaganda film, and again afterward. A control group fills out the scale twice without seeing the film. The control group is used to measure practice effects. The change in the scores of the experimental group relative to the control group, whether their attitudes have become more or less favorable, indicates the effects of the film.

Thurstone-type Scale

In 1928 L. L. Thurstone published the article "Attitudes Can Be Measured." The title was radical because many psychologists of his day doubted that something as subjective and vague as an attitude could be measured. Thurstone maintained that people's attitudes on a topic represented the sum total of their beliefs, feelings, knowledge, and opinions about it. He developed an elaborate procedure for measuring attitudes.

Constructing a Thurstone-type Scale

The first step in constructing a Thurstone-type scale is to collect statements on the topic from people holding a wide range of attitudes, from extremely favorable to extremely unfavorable.

Examples of attitudes toward the use of marijuana:

It has its place.
Its use by an individual could be the beginning of a sad situation.
It is perfectly healthy; it should be legalized.

Duplications and irrelevant statements are omitted. The rest are typed on 3 × 5 cards and given to a group of people who serve as judges. The judges need not have any special qualifications other than to be representative of the population whose attitudes you wish to measure. They are asked to sort the statements into eleven piles representing the entire range of attitudes from extremely unfavorable (1) to extremely favorable (11). Pile 6 is for statements that are neither favorable nor unfavorable. The location of the pile in the series indicates the judges' ratings of the item (pile 8 indicates a rating of 8). Thus each statement will receive a rating from 1 to 11 from each judge. The ratings from each statement are averaged, and that average indicates how favorable the statement is on the topic.

Example

	Average rating from 20 judges (11 = extremely favorable)
If marijuana is taken safely, its effects can be quite enjoyable.	8.9
I think it is horrible and corrupting.	1.6
It is usually the drug people start on before addiction.	4.9

Note that the judges do not indicate their personal opinion on the issue; they simply decide how favorable or unfavorable each statement is. Statements about which the judges are unable to agree are not used in the final scale. The statement "Marijuana use should be taxed heavily" was rejected because it was ambiguous. Some judges thought it was pro-marijuana because it implied legalization, while others felt it was anti-marijuana because it advocated a heavy tax.

Administration and Scoring

After the statements have received their average ratings from the judges, they are arranged in random order on a questionnaire. Table 10-1 illustrates a Thurstone-type scale for measuring attitudes toward marijuana use. The

TABLE 10-1. Thurstone-type Scale for Measuring Attitudes toward Marijuana

This is a scale to measure your attitude toward marijuana. It does *not* deal with any other drug, so please consider that the items pertain to marijuana exclusively.

We want to know how students feel about this topic. In order to get honest answers, the questionnaires are to be filled out anonymously. Do *not* sign your name.

Please check all those statements with which you agree.

___ 1. I don't approve of something that puts you out of normal state of mind. (3.0)*
___ 2. It has its place. (2.2)
___ 3. It corrupts the individual. (2.2)
___ 4. Marijuana does some people a lot of good. (7.9)
___ 5. Having never tried marijuana, I can't say what effects it would have. (6.0)
___ 6. If marijuana is taken safely, its effects can be quite enjoyable. (8.9)
___ 7. I think it is horrible and corrupting. (1.6)
___ 8. It is usually the drug people start on before addiction. (4.9)
___ 9. It is perfectly healthy and should be legalized. (10.0)
___10. Its use by an individual could be the beginning of a sad situation. (4.1)

*Average ratings from statements sorted into 11 piles, from very unfavorable (1) to very favorable (11). These numbers were *not* included in the actual scale to be filled out by respondents.

average ratings made by the judges are shown in parentheses. These should not actually be reproduced on the final questionnaire.

People filling out a Thurstone-type scale are asked to check only those statements with which they agree. No indication of the strength of agreement is given. The respondent's score is computed by adding up the average ratings of all the statements checked and dividing by the number of statements checked.

Example: Scoring a Thurstone-type scale
Favorability rating (number in parentheses) of each statement checked.
2.2
1.6
4.9
Total: 8.7 ÷ 3 (number of statements checked) = 2.9 (score)

Likert Scale

The Thurstone procedure for constructing an attitude scale was later simplified by Likert (1932).

Constructing a Likert-type Scale

A Likert-type scale includes only statements that are clearly favorable or clearly unfavorable. Statements that are neutral or borderline are eliminated. Judges are used in making this decision. They rate each statement as to whether it is favorable or unfavorable on the topic. Where there is little agreement among the judges or difficulty in deciding whether the item is favorable or unfavorable, the statement is eliminated. Rather than using precise mathematical averages for each degree of favorability, Likert recommended that all items be given the same mathematical weight.

The Likert procedure eliminates the need for establishing precise favorability weights (scores) for each statement. It also simplifies the scoring procedure by using whole numbers (e.g., 1 to 5) for each item rather than numerical averages (e.g., 2.8 or 3.6). A Likert-type scale presents a list of statements on an issue to which the respondent indicates degree of agreement using categories such as Strongly Agree, Agree, Undecided, Disagree, and Strongly Disagree.

Administration and Scoring

The statements are arranged in random order on a questionnaire with a choice of degrees of agreement. Each favorable statement would be followed by five degrees of agreement (strongly agree, agree slightly, undecided, disagree slightly, strongly disagree), which would be scored 5, 4, 3, 2, and 1, respectively. The same procedure would be done with unfavorable statements except that these would be scored in the reverse direction (1, 2, 3, 4, and 5, respectively). Respondents are asked to check their degree of agreement or disagreement with all statements on the list.

Example

Indicate your degree of agreement or disagreement with each of the following statements about marijuana by circling one of the following letters

SA = Strongly Agree
A = Agree
U = Undecided
D = Disagree
SD = Strongly Disagree

SA A U D SD 1. Marijuana use corrupts the individual.
SA A U D SD 2. Its use by the individual could be the beginning of a sad situation.
SA A U D SD 3. Marijuana does some people a lot of good.

People with very favorable attitudes would be expected to strongly agree with the favorable statements and strongly disagree with the unfavorable statements. They would earn a high score on the scale when the item scores are added together. Conversely, people with very unfavorable attitudes would be expected to strongly disagree with the favorable statements and strongly agree with the unfavorable statements, and would score low on the scale. The Likert and Thurstone procedures for constructing attitude scales are compared in Table 10-2. The simplicity of the Likert scoring system

TABLE 10-2. Comparison of the Thurstone and Likert Procedures for Constructing Attitude Scales

Thurstone procedure	*Likert procedure*
Statements collected from a wide range of individuals	Same
Statements sorted by judges into piles representing 11 degrees of favorability on the topic	Statements rated by judges as favorable, unfavorable, or neutral on a topic
Statement on which judges disagree as to favorability discarded	Same
Items selected to represent the full range of opinion, from very favorable to very unfavorable, including neutral statements	Neutral statement discarded, leaving only statements clearly pro or con Each statement followed by the terms "strongly agree," "agree," "undecided," "disagree," "strongly disagree"
Respondents asked to check only those statements with which they agree	Respondents asked to indicate their degree of agreement or disagreement with all statements
Each statement scored according to average degree of favorability (e.g., 2.1, 4.6)	Favorable statements scored from 5 to 1; unfavorable statements scored from 1 to 5
Mathematical average of all statements checked is computed	Scores for all items are summed

makes this procedure attractive to researchers. Most attitude scales used today are of the Likert type.

Validity and Reliability

The *validity* of an attitude scale is the degree to which it measures the attitude or belief system in question. A common method for assessing validity is to administer the attitude scale to individuals known to hold strong opinions on both sides of an issue. For example, a scale measuring tolerance for minorities could be administered to members of a segregationist group and to members of an organization working for racial equality. If the scale is valid, there will be a large difference between the responses of the two groups. An attitude scale should yield consistent results. Consistency in measurement is known as *reliability.* There are three common methods for testing the reliability of an attitude scale: test-retest, split-half, and equivalent forms. With the *test-retest* method, the scale is given to the same person on two occasions and the results are compared. Unless something significant happened during the interval, the two scores should be similar. The *split-*

half method involves splitting an attitude scale into two halves, which are then compared. This is generally done by combining all the even-numbered items into one scale and all the odd-numbered items into another. The scores on the two halves are compared and should be similar if the scale is reliable. The third method of measuring reliability involves the use of *equivalent forms.* Two different scales on the topic are constructed, Form A and Form B. If the scale is reliable, the scores on the two forms should be similar.

These three methods for determining reliability rest on a comparison between two sets of scores. This comparison is made through a statistical test known as the correlation coefficient, described in Chapter 18.

For readers who do not want to construct their own attitude scales, a good selection of scales whose reliability has already been established is available in Robinson and Shaver, *Measures of Social Psychological Attitudes* (1973). Antonak and Livneah (1988) have assembled 22 scales assessing attitudes toward the disabled, including information on the reliability and validity of each scale. Journal articles on the topic of interest will also provide references to scales currently in use.

Limitations

There are serious questions about the validity of attitude scales. Often they predict behavior poorly or not at all. The words on the printed page bear little resemblance to the actual situation. Another problem with attitude scales is the assumption that attitudes lie along a single dimension of favorability. People's opinions on a topic like marijuana are complex and multidimensional. A person may be in favor of reducing the penalties on marijuana posession but not on cultivation or sale, and may want strict penalties for anyone driving under the drug's influence. A single favorability score cannot reflect the specificity of these concerns. A questionnaire, which treats each item separately, is more widely used than the attitude scale for this reason.

Semantic Differential

The semantic differential is a procedure developed by psychologist Charles Osgood and his associates to measure the meaning of concepts (Osgood, May, & Miron, 1975). The respondent is asked to rate an object or a concept along a series of scales with opposed adjectives at either end.

Example

Rate the park nearest your residence along each of the following scales.

Quiet ___:___:___:___:___:___:___ Noisy

Dangerous ___:___:___:___:___:___:___ Safe

Sad ___:___:___:___:___:___:___ Happy

The semantic differential is a good instrument for exploring the *connotative meaning* of things. Connotation refers to the personal meaning of something, as distinct from its physical characteristics. For example, a panther, in addition to being a large cat, connotes stealth and power. Crêpes Suzette suggest elegance and expensive dining.

Selection of Terms

In the research that developed the semantic differential, three major categories of connotative meaning were found: value (e.g., good–bad, ugly–beautiful), activity (e.g., fast–slow, active–passive), and strength (e.g., weak–strong, large–small). Table 10-3 presents four adjective pairs high in value, activity, or strength. Not surprisingly, the value dimension (good–bad, valuable–worthless) is of greatest importance in evaluative research. When you want to know whether or not people like something, you will probably want to include good–bad, ugly–beautiful, and friendly–unfriendly. Activity and strength are important dimensions in certain circumstances. A comparison of people's images of cities and small towns found major differences on the activity and strength dimensions. Cities were full of bustle, hurry, and activity, while in small towns the pace was more slow, relaxed, and leisurely. Cities were also rated as larger, stronger, and more powerful than small towns. Other adjectives may be more relevant to a particular topic. An investigation of religious concepts used adjectives closely related to religious belief, such as sacred–profane, mysterious–obvious, and public–private. The nature of the project will determine the adjectives to be used.

The most common error made by inexperienced researchers using this technique is to overestimate the respondents' vocabulary level. Although most college students know the meaning of "profane" and "despotic," a substantial number may not, which reduces the validity of the results when these terms are included on a rating scale. Pretesting the adjective pairs is essential for eliminating difficult or ambiguous terms. Even if adjectives have been used by other researchers, it will still be necessary to test them on your particular respondents. Adjectives that have one meaning for one group of people may mean something else to another group.

TABLE 10-3. Polar Adjective Pair*

High on value dimension	*High on activity dimension*	*High on strength dimension*
Good–bad	Active–passive	Strong–weak
Beautiful–ugly	Energetic–inert	Large–small
Friendly–unfriendly	Fast–slow	Hard–soft
Wise–foolish	Excitable–calm	Heavy–light

*Adapted from Osgood, Miron, & May, 1975.

Length and Layout

Don't burden your respondents with too many scales. After a while, the lines become a blur. We do not recommend using more than 20 adjective pairs to measure a concept; 10 to 12 adjective pairs seem preferable. Remember that the value of your results depends on the voluntary cooperation of your respondents.

Counterbalance the order of positive and negative adjectives. Begin some scales with the positive term (happy–sad) and others with the negative term (noisy–quiet). This will prevent the respondent from falling into a fixed pattern of always checking to the right or left.

Make sure that people put their marks in the right place. Most researchers use solid lines for the responses and colons as spacers.

Example:

Light ___:___:___:___:___:___:___ Dark

It is important that answers be marked on the lines and *not* on the dots. Tabulating the responses becomes very complicated when people have checked on the dots. When this occurs, many researchers assign the mark to the right or left line in random order. That is, if a person has checked midway between the second and third line, the response will be scored as a 2 the first time and a 3 the next time this occurs. So long as this is done randomly, the results should not be biased.

Most researchers follow Osgood in using seven-point scales. This includes a midpoint, which is useful when the item is neither happy nor sad or neither light nor dark, but somewhere in the middle. However, if machine scoring limited to a five-point scale can be done cheaply and quickly, this option should be seriously considered. Five-point scales are more easily tabulated by hand, too. Many researchers find that differences among the three scale points to the right or left of the midpoint have little meaning. The direction of response (e.g., whether the cafeteria is seen as a happy place) is more important than whether it is seen as extremely happy, moderately happy, or somewhat happy. If you plan to combine all three categories to the right of the neutral point later, you might as well begin with a smaller number of scale points—five or even three.

Example

Rate this room along each of the following scales:

friendly ___:___:___ unfriendly
light ___:___:___ dark
quiet ___:___:___ noisy
dirty ___:___:___ clean

Scoring

The scale points are given numerical values from 0 to 6 or 1 to 7 on a seven-point scale, going from left to right. The average is computed separately for each pair. Thus, 3 is the midpoint value of the happy–sad scale, whose end points are 0 and 6. Anything below 3 means that the item is generally happy, and anything above 4 means that the item is generally sad. In presenting the results, it is helpful to the reader to reorganize all the scales so that the favorable end is on the left and the unfavorable end on the right. Note that this differs from the order of the scales given to the respondents. Placing all the favorable adjectives on the left enables the reader to see at a glance how the ratings came out.

The results can be presented graphically as well as in averages. Figure 10-1 shows student ratings of a reading room in a university library. The room is highly valued and strong but relatively low in activity.

Limitations

The semantic differential is usable only with intelligent and cooperative adults. People with little education often focus on the ends of the scale and do not use the middle points. We would not recommend using the semantic differential with children, with people whose command of the language is limited, with older people who would have difficulty seeing the various scale points, or with any group of respondents who are not accustomed to making fine distinctions in their assessments of persons, objects, and events.

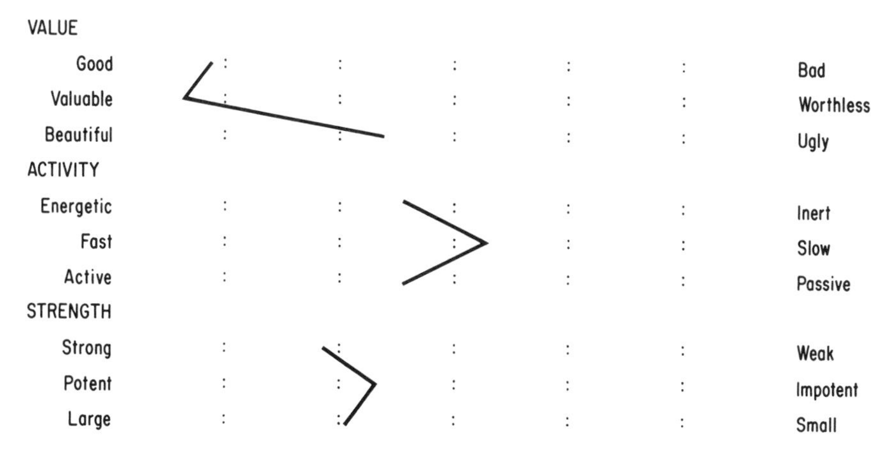

FIG. 10-1. Library reference room, ratings from the semantic differential.

Performance Rating Scales

Many companies require the work of all employees to be formally rated periodically. This is intended to ensure some level of competence among workers. It also requires a supervisor to have firsthand contact with the employee's work. Performance scales contain a built-in evaluative component. Behavior is rated along dimensions such as skilled–unskilled, organized–disorganized, competent–incompetent. It is desirable to make the behaviors assessed very specific; instead of asking a general question about work habits, the rater is asked about the employee's response to supervision, ability to work under pressure, relationships with fellow employees, and so on. These items can be collected by asking workers and supervisors about what qualities are important on the job and from previous research on worker productivity. It is important to pretest both the job-related characteristics and the rating scales to eliminate ambiguous or unclear terms.

Experience with performance ratings in industry, the civil service, and the armed forces has been disappointing. These ratings are still used in many settings primarily because no better measures are available. *Halo effects* refer to the tendency to make ratings of specific abilities on the basis of an overall impression. If you like someone, it is easy to see everything he does as good. If you don't like him, it is easy to give poor marks to all his work. Another problem is that raters are not willing to make honest judgments. There are severe pressures against saying unkind things about another person. If a supervisor wants to be known as a "good guy," all employees routinely get good evaluations. Also, low ratings for employees reflect badly on the supervisor.

Because raters know that extremely positive statements are the rule, anything less than the top score is seen as likely to jeopardize the person's career. With a shrug of the shoulders because "everybody does it," the rater routinely checks "outstanding" for all categories. This happened in the U.S. Army Officer Efficiency Ratings. In this scale, "Excellent" was a typical score and "Very Good" became a mark of disapproval (Thorndike, 1962). This tendency to inflate a person's abilities is readily apparent on recommendation forms that use performance scales. A high proportion of recommendation letters describe the applicant as being in the top 5 or 1 percent.

One method used to minimize the tendency to give top ratings to everyone is to make the questions specific and personal. Instead of asking whether an employee is a good worker, supervisors are asked how likely they would be to recommend the employee for advancement in their department. A professor writing a letter of recommendation would be asked not only whether the applicant was a good student but whether the professor would admit the student to his or her own graduate program. Personalizing the questions helps reduce the ho-hum attitude on the part of the rater. The

BOX 10-1. Air Force Ends Quota on Top Ratings

As of October, 1978, the Air Force will no longer impose a quota on the number of officers who can receive the highest efficiency ratings. The quota had been imposed in 1975 when it was found that 95% of officers were receiving "ones" which was the highest possible performance rating. Air Force officials decreed that only 22% of the officers within a command could receive ratings of "1" and only 28% could receive the lesser rating of "2." The remainder would have to receive lower ratings from 3 to 6.

Outraged responses from those doing the rating and those being rated forced the Air Force to withdraw the quotas. Critics charged that the quota system did not take into account the quality of officers on a base. In a mediocre command, even average performance could rate a "1" but an outstanding performance on a top notch base might not even rate a "2." The Air Force bowed to the criticism and returned to the older system of permitting an unlimited number of top scores (Bell, 1978).

generosity factor may be less apparent in anonymous ratings. While anonymity for raters is impractical in considering applicants for admission or promotion, it is quite feasible in research.

Limitations

Performance rating scales have not been very useful in research because of the reluctance of raters to say unkind things about people, halo effects, and a lack of standards for judging employee effectiveness. If other criteria of effectiveness are available, such as production records or customer ratings, the supervisor's rating may provide useful supplementary information.

Consumer Rating Scales

When checking into a motel room, it is common to find a short questionnaire on the dresser asking for an evaluation of the service, facilities, and food. The purchaser of a new car is likely to receive a questionnaire in the mail from the national distributor asking about the quality of dealer service and maintenance. Consumer organizations collect evaluations of products from members and volunteers. Such ratings of actual products and services are different from market surveys to establish levels of demand. Rating is most appropriate for items with which the person is directly familiar. Surveys of needs and wants are better handled through interviews.

Product rating scales are useful for comparing responses from large groups. It would be awkward and time-consuming to compare interviews collected from 50 owners of car A and car B. Comparison is easier if all persons rate their cars along the same set of scales. Instead of using a random or representative sample of the community, it is common to use only those who have had direct experience with the product or service. Thus, ratings of an airline would be obtained from passengers during a flight. A questionnaire regarding customer satisfaction with a camera might be attached to the warranty card. Commercial firms employ methods that vary in complexity and sophistication to obtain customer feedback. Some cast their nets widely in the hope of finding something useful; others prefer more detailed surveys with a much smaller, carefully selected sample. Often the latter surveys are undertaken by a national marketing firm.

The first step in product rating is to identify those characteristics that are important and relevant. This is done by examining ratings of similar products and by consulting with suppliers and customers.

The next step is to establish scale points. For a brief questionnaire that accompanies the product or is filled out by customers, a three-point scale is probably sufficient.

Example
Please tell us how you felt about your stay in this hotel.

desk clerk	above average	average	below average
elevator	above average	average	below average
cleanliness of room	above average	average	below average
coffee shop service	above average	average	below average

For children and others not accustomed to making verbal ratings, a series of facial expressions can be used to indicate liking.

Example
Which face shows the way you feel about this product?

Limitations

Ratings through questionnaires attached to the product or left on hotel dressers are subject to response bias. Persons most likely to fill out and send in questionnaires will be those with strong opinions pro and con—and generally the latter. Response rates will vary with the consumers' interest in helping the manufacturer or service agent.

Sensory Evaluation

Sensory evaluation began in the laboratories of early experimental psychologists, who were interested in the basic properties of odors, tastes, sound and other sensations. The connection between the physical qualities of objects and their sensory attributes is called *psychophysics.* A key assumption in psychophysics is that people can make meaningful ratings of the degree of their sensory experiences (e.g., rating items as more or less bright, loud, sweet, and so on).

The food and beverage industries rely heavily on sensory evaluation. Before a new product is marketed, its consumer acceptance will be tested. Products are often first tested by expert judges who have exceptionally well-developed palates, noses, or visual sensitivity before being tried out on a panel of nonexperts. Researchers in Norway examined consumer response to black currant juice, which varied in strength, color, acidity, portion size, and time of testing (before or after lunch). Preference was found to be mainly influenced by color, acidity, and portion size (Martens, Risvik, & Schutz, 1983).

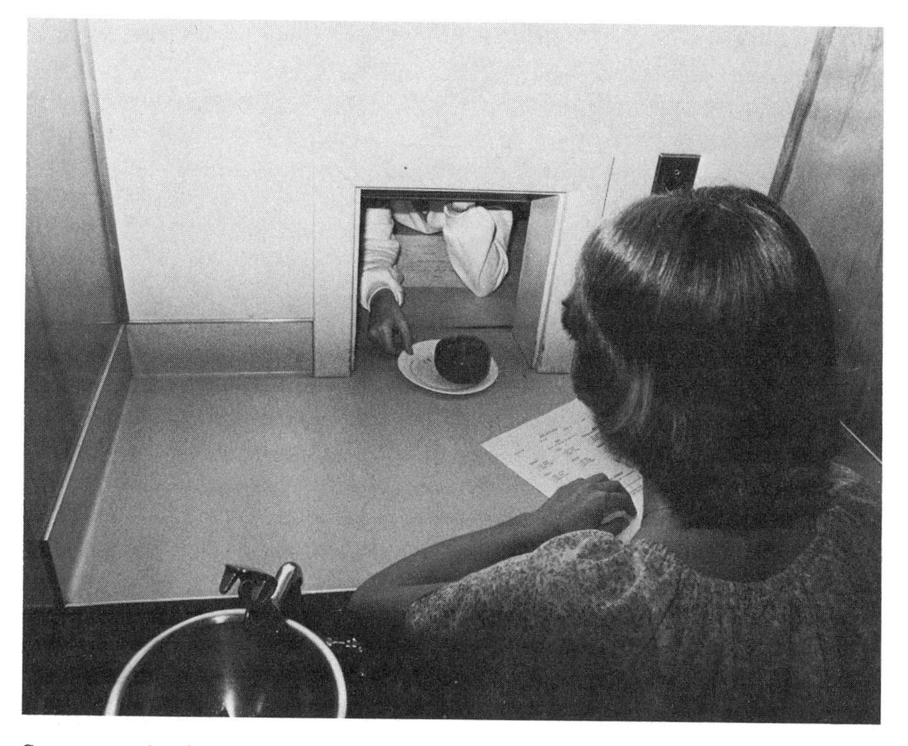

Sensory evaluation. The student was asked to rate the flavor and appearance of tomatoes.

The qualities to be rated depend as much on the interests of the investigator as on the objective characteristic of the item. A firm might be interested in the visual appearance of a bar of soap, the texture of canned fruit, or the sound level of fluorescent lights. Deciding what characteristics are relevant should be done in consultation with the client or consumer organization, or it can be based on previous research.

Various methods have been used to present material to the judges. One approach is to present the judges, at the beginning of the session, with *standards*. For an investigation of taste qualities, the judge will first taste four different compounds, one very sweet, one very sour, one very salty, and another very bitter to use as standards in making subsequent judgments.

Example

Rate the item you have tasted along each of the following scales. Place a check anywhere along the line.

very sweet	not sweet
very sour	not sour
very salty	not salty
very bitter	not bitter

Note that the four taste qualities are rated separately. Sweet is not considered the opposite of sour. Grapefruit and pineapple can be both sweet and sour.

In the method of *paired comparisons,* items are presented two at a time and the person is asked to compare them. This method is useful in deciding whether a change represents an improvement over a standard model.

Example

Compared to B (the standard), item A is:

spicier	less spicy	equal
sweeter	less sweet	equal
more dense	less dense	equal
more fragrant	less fragrant	equal

Since the subject compares only two items at a time, each comparison can be done quickly and easily. There is very little dependence on memory. Comparison procedures are useful with inexperienced judges, who can express a preference for one item over another without being specific as to their reasons.

Example

Which of these two wines is sweeter, A or B?
Which of these two wines would you choose to accompany a steak dinner, A or B?

Such sessions are conducted as *blind taste trials*. The term *blind* indicates that the subject is not aware of the origin or identity of the item being rated. The subject is told its general category (wine) but not the specific variety, cost, or place of origin. Blind tasting minimizes the effects of labels and stereotypes. Subjects may be more likely to give high ratings to wines with expensive labels or fancy names. A further refinement of this procedure requires two experimenters, one who replaces all identifying information with code numbers before the sessions. The second experimenter, who has no information on the coding system, conducts the actual taste trials. This is called a *double blind procedure,* since both the subject and experimenter conducting the tests are in the dark about what is being tasted.

Limitations

Like performance rating, sensory evaluation is subject to a halo effect. When people like a product, they tend to see most things about it as good; if they dislike it, they see everything about it as bad. Without careful explanation, the terms used in sensory evaluation may not be clear to those doing the rating; for example, people may have difficulty distinguishing among fragrant, fruity, and spicy. Expert judges, such as food critics and wine tasters, use different criteria than those used by ordinary consumers. Sensory evaluation requires people to make artificial distinctions. When they taste ketchup on a hot dog, most people do not divide the taste into separate degrees of sweetness, sourness, and saltiness.

Summary

Rating scales are used to rank people's judgments of objects, events, or other people from low to high or from good to poor. They provide numerical scores that can be used to compare individuals and groups.

On a graphic rating scale, the respondent places a mark along a continuous line. On a step scale, the rater checks one of a graded series of steps without intermediate points. On a comparative rating scale, the person is asked to compare the object or person with others in the same category.

Four levels of measurement are nominal, ordinal, interval, and ratio. The level of measurement determines the types of comparisons that can be made within a scale, and the statistics used.

An attitude scale is a special type of questionnaire designed to produce scores indicating the overall degree of favorability of a person's attitudes on a topic. A Thurstone-type scale contains statements expressing a full range of opinions on a topic. Respondents check all those statements with which they agree. A Likert-type scale contains only statements that are clearly favorable or clearly unfavorable. No neutral or borderline statements are included. The respondents rate each statement along a five-point scale of agreement, from strongly agree to strongly disagree. Most attitude scales used today are of the Likert type.

Reliability refers to consistency of measurement. There are three common methods for estimating the reliability of an attitude scale. In the test-retest method, the scale is given to the person on two occasions and the results are compared. The split-half method involves splitting an attitude scale into two halves which are then compared. The third method of measuring reliability involves constructing two equivalent forms of the scale. If the scale is reliable, the person's score on the two forms should be similar. The chief limitation of attitude scales is that they predict behavior poorly or not at all.

The semantic differential is a procedure developed to measure the connotative meaning of concepts. Connotation refers to the personal meaning of something as distinct from its physical characteristics. Three major categories of connotative meaning are value, strength, and activity.

Performance rating scales are used to judge the competence or work output of employees. Experience with their use in most settings has been disappointing. Many supervisors are not willing to make honest judgments. The halo effect refers to the tendency to rate specific abilities on the basis of an overall impression.

Consumer ratings are used to find out people's opinions about products and services with which they are familiar.

Sensory evaluation is used to test the psychophysical properties of products, particularly food and beverages. Sensory evaluation can involve expert judges or average persons. Sometimes judges are asked to rate items along graphic rating scales (e.g., sweet–not sweet, salty–not salty). In the method of paired comparisons, items are presented two at a time and the person is asked to compare them. In a blind taste trial, the respondent does not know the origin or specific identity of the item being rated. In a double-blind procedure, neither the subject nor the experimenter knows the origin or specific identity of the item being rated.

Without careful explanation, the terms used in sensory evaluation may not be clear to those doing the rating. Expert judges such as food critics use different criteria than those used by ordinary consumers.

References

Antonak, R. F., & Livneah, H. (1988). *The measurement of attitudes toward people with disabilities: Methods, psychometrics and scales.* Springfield, IL: Charles C. Thomas.

Bell, T. (1978, October 17). Air Force ends quota on ratings. *Sacramento Bee,* p. B3.

Likert, R. (1932). A technique for the measurement of attitudes. *Archives of Psychology, 140,* 1–55.
Martens, M., Risvik, E., & Schutz, H. G. (1983). Factors influencing preference: A study on black currant juice. *Proceedings of the Sixth International Congress of Food Science and Technology,* 2, 193–194.
Osgood, C. E., May, W. H., & Miron, M. S. (1975). *Cross-cultural universals of affective meaning.* Urbana, IL: University of Illinois Press.
Robinson, J. P., & Shaver, P. R. (1973). *Measures of social psychological attitudes* (rev. ed.). Ann Arbor, MI: Survey Research Center, Institute for Social Research, University of Michigan.
Thorndike, R. L. (1962). *Personnel selection.* New York: Wiley.
Thurstone, L. L. (1928). Attitudes can be measured. *American Journal of Sociology, 33,* 529–554.

Further Reading

Anderson, A. B., Basilevsky, A., & Hum, D. P. J. (1983). Measurement: Theory and techniques. In P. H. Rossi, J. D. Wright & A. B. Anderson (Eds.), *Handbook of survey research* (pp. 231–287). New York: Academic Press.
Dunn-Rankin, P. (1983). *Scaling methods.* Hillsdale, NJ: Erlbaum.
Henerson, M. E., Morris, L. L., & Fitz-Gibbon, C. T. (1987). *How to measure attitudes.* Beverly Hills, CA: Sage.
Mueller, D. J. (1986). *Measuring social attitudes: A handbook for researchers and practitioners.* New York: Teachers College Press.
Robinson, J. P., & Shaver, P. R. (1973). *Measures of social psychological attitudes* (rev. ed.). Ann Arbor, MI: Survey Research Center, Institute for Social Research, University of Michigan.
Schuman, H., & Presser, S. (1981). *Questions and answers in attitude surveys.* New York: Academic Press.

V

Other Useful Research Procedures and Materials

11 Content Analysis

Controlling False Advertising

Two consumer researchers were interested in seeing how the regulatory climate in Washington, D.C., affected the product claims made in advertisements. They selected three time periods for their study. In 1970 there was minimal regulation of advertising by the Federal Trade Commission. In 1976 there was a spurt of activism by the agency in pressing manufacturers to reduce false claims. This was followed by a decade of deregulation when the agency became more passive. The researchers hypothesized that the number of advertising claims would be lowest during the period of regulatory activism, and that the effect would be most pronounced in those industries that had been required by the government to provide proof of their advertising claims.

The first task of the researchers was to select consumer products on which to test their hypotheses. They searched through government records to find items that had been the subject of government orders requiring manufacturers to verify their claims. They also sought products heavily advertised in popular magazines. From the list of items meeting these two criteria (government orders and heavy advertising), two were selected: antiperspirants and pet foods. To serve as a basis of comparison ("control group" of products), a list was made of items that had not been challenged by the government, but which were also advertised heavily in popular magazines. Skin lotion and prepared foods were chosen from this list.

The next task was to collect the magazines in which the ads would be studied. The researchers began with a list of American consumer magazines with circulations above 500,000. Those with few or no ads in the product categories were eliminated. From the remaining magazines, sixteen were randomly selected, including *Better Homes and Gardens, Glamour, News-*

week, Reader's Digest, and *Sports Illustrated.* Color reproductions were made of the advertisements, which were classified by university students recruited by the researchers. Each student was given approximately 20 ads to classify. Of the 662 ads, approximately one third were seen by at least two students in order to determine reliability. Categories used in the analysis were *product claim,* the degree to which the ad made a claim about effects, and the amount of *verification* (proof) contained in the ad.

Reducing the various findings to the key issues, the authors concluded that government regulation had an impact on advertising claims. Active regulation reduced the amount of information given to consumers, but it was better quality information (Kassarjian & Kassarjian, 1988).

Quantification in Content Analysis

Content analysis is a technique for systematically describing the form and content of written or spoken material. It has been used most often in the study of mass media. However, the technique is suitable for any kind of written or spoken material, including interview records, letters, songs, cartoons, advertising circulars, and so on. The basis of a content analysis is *quantification* (i.e., expressing data in numbers). Instead of impressions about trends and biases, the investigator comes up with precise figures. A content analysis of medical advertising found that 96 percent of the models used for physicians were male, while 90 percent of the nurses or lab technicians were female. These figures strengthen the case for a stereotyped role portrayal and allow comparison with other media.

The analysis can emphasize the content of the material or its structure or both. *Content* refers to the specific topics or themes. *Structure* includes location on the page, format, use of illustrations, and so on. How much attention does a president's speech devote to the Soviet Union compared to China? This is basically a structural question. Whether the speech is anti-Soviet or pro-Soviet, anti-Chinese or pro-Chinese, is a content question.

Content analysis allows a person to do social research without coming into contact with people. There is no need to check with a human subjects committee. No laboratory equipment or expensive facilities are required. The materials for a content analysis are probably available in a library or newspaper file. In some cases, it will be necessary to dig for scarce materials in back issues of a periodical that is difficult to locate. A reference librarian can be extraordinarily helpful in finding materials.

Content analysis has additional advantages when compared with other techniques. The investigator can use material that is already available. This eliminates the possible bias that occurs when the researcher generates the data in an interview or observational study. Content analysis is unobtrusive. The observer has no effect upon the material collected. The material is the same after the study as before. This is not true in an experimental or interview study. Content analysis permits the comparison of trends over time. It is easy for one person to repeat another's study since the materials are available and unchanged by the previous study. The study of product advertising described at the beginning of this chapter was done in two phases. The first phase compared advertising for the years 1970 and 1976 (Healey & Kassarjian, 1983); the follow-up study conducted 8 years later compared 1984 advertising with the earlier results (Kassarjian & Kassarjian, 1988). The existence of this data base makes it likely that further comparisons will be made to test additional hypotheses. Of all the methods discussed in this book, content analysis scores the highest marks in terms of ease of replication. The technique is well suited to comparisons among different nations. The method also allows for the simultaneous application of quantitative and qualitative techniques. The structural aspect of advertising or news reports can be counted. At the same time the content can be classified qualitatively according to the values or attitudes expressed.

Although many researchers use this technique with material already printed, it is also possible to generate new material for a content analysis. An Israeli psychologist performed a content analysis on written reports of dreams collected from Jewish and Arab fifth-to-seventh graders living in various parts of Israel and the West Bank (Bilu, 1989). Each dream was analyzed for the degree to which Jews appeared in Arab dreams and vice versa. There was also an analysis of instances of friendliness and aggression between the dream characters. In this case, the researcher did "react" with the participants in the sense of requesting them to write down their dreams.

On the debit side, content analysis can be a tedious and exacting activity. Imagine sitting at a library table with 10 years of magazines stacked in front of you. Each article must be classified along several dimensions. The first few issues are interesting. By the time the twelfth issue is done, it is easy to wonder why you embarked on the study. Content analysis requires patience and attention to detail. It is not a technique for someone who wants to determine the main trends and run. It requires long hours at a library or study table counting, measuring, and classifying. Yet there is an excitement that comes from seeing trends emerge from a mass of printed material. A researcher was interested in book acknowledgments, which are statements at the beginning of a book in which people give credit to others. He wanted to know whether gender played a role in the likelihood that someone would be credited. When the person acknowledged was a spouse, women generally thanked their husbands for advice, while men thanked their wives mostly for typing (Moore, 1984).

Finding Categories

The best way to select categories for classification is first to skim over the materials to identify the major themes. These can be listed as they are found. When the categories begin to repeat themselves (i.e., the material all seems to be covered by the previous categories), you can stop for the moment.

Categories that overlap or duplicate one another can be combined. The resulting categories can then be tried on new materials to see how well they fit.

The list of categories must be comprehensive, covering all the items to be analyzed. However, it is legitimate to use an "other" heading for unusual items that belong nowhere else and aren't sufficiently numerous to warrant a separate heading.

Example

Content analysis of responses to open-ended interview question: "What is the single most important reason why you shop in this store?"

Reason	*Classification*
"Prices are low"	Prices
"Near my home"	Location
"Friendly service"	Employees
"Prices cheap"	Prices
"Convenient"	Location
"Good value"	Prices
"Accompanied friend"	Other
"Store on my way to work"	Location

Establishing categories is not always easy. Cartoons may be difficult to classify with respect to type of humor. Some cartoons may show a combination of categories (e.g., both aggressive and sexual or absurd and sick). Classifying emotional themes in television serials will have a subjective component. In general, structural items such as amount of space, time, and location tend to be more reliable than content themes, which are more subjective.

When coding specific elements, it is important to record noninstances too. If you are going to count the number of women with gray hair shown on TV, you should also count all the women shown on the screen as a basis for comparison. The categories are placed on a coding sheet to be used by the coders, those individuals who will classify the content and structure of the material into specific categories. Box 11-1 shows the coding system used in a study of medical advertisements.

BOX 11-1. Coding System for Medical Advertisements

Ad sheet (Fill out one for each advertisement)

Column	*Variable code*
1	Journal (circle one)
	1. *American Journal of Medicine*
	2. *JAMA*
	3. *Medical Aspects of Human Sexuality*
	4. *Psychiatric News*

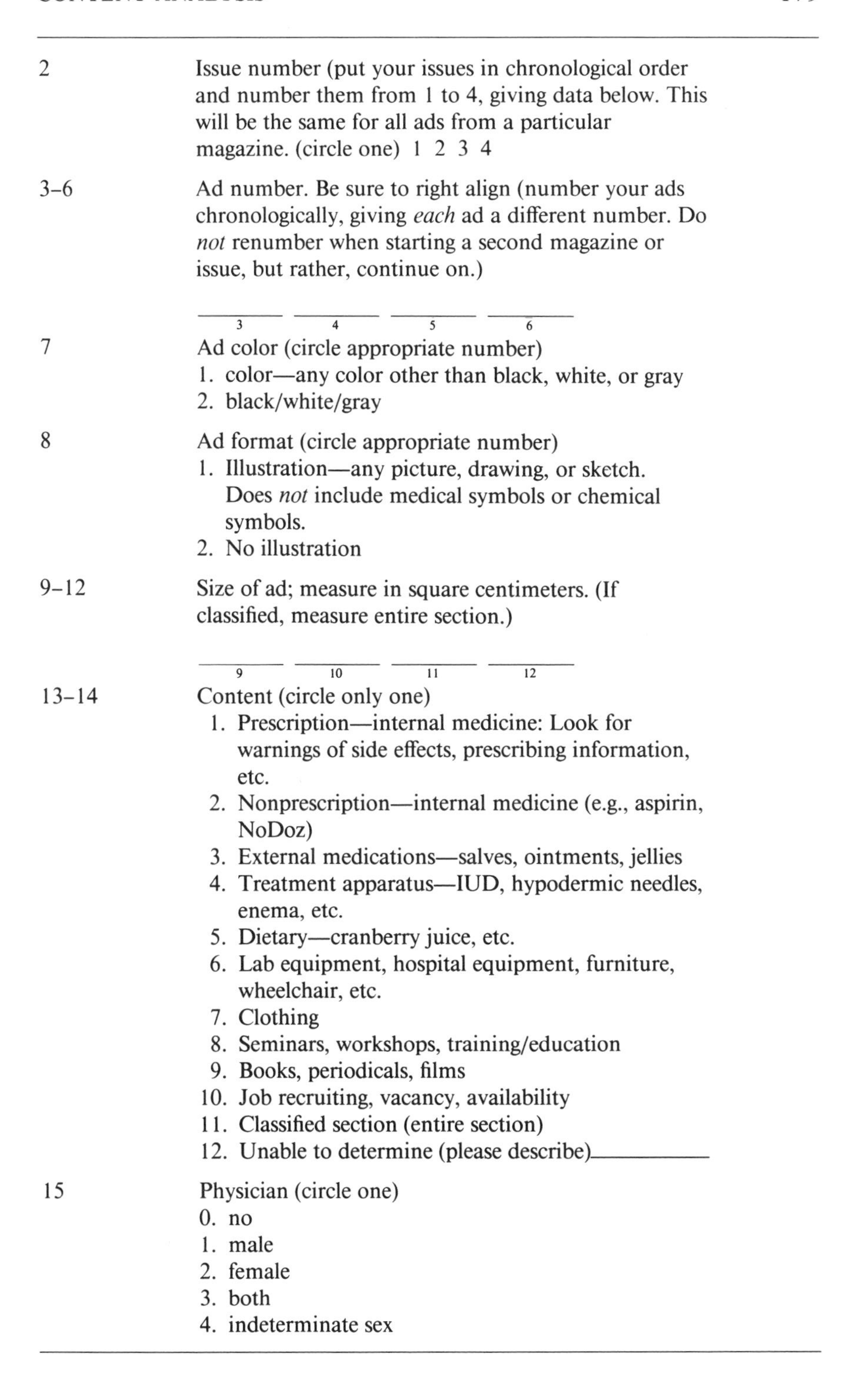

2 — Issue number (put your issues in chronological order and number them from 1 to 4, giving data below. This will be the same for all ads from a particular magazine. (circle one) 1 2 3 4

3–6 — Ad number. Be sure to right align (number your ads chronologically, giving *each* ad a different number. Do *not* renumber when starting a second magazine or issue, but rather, continue on.)

____ ____ ____ ____
3 4 5 6

7 — Ad color (circle appropriate number)
1. color—any color other than black, white, or gray
2. black/white/gray

8 — Ad format (circle appropriate number)
1. Illustration—any picture, drawing, or sketch. Does *not* include medical symbols or chemical symbols.
2. No illustration

9–12 — Size of ad; measure in square centimeters. (If classified, measure entire section.)

____ ____ ____ ____
9 10 11 12

13–14 — Content (circle only one)
1. Prescription—internal medicine: Look for warnings of side effects, prescribing information, etc.
2. Nonprescription—internal medicine (e.g., aspirin, NoDoz)
3. External medications—salves, ointments, jellies
4. Treatment apparatus—IUD, hypodermic needles, enema, etc.
5. Dietary—cranberry juice, etc.
6. Lab equipment, hospital equipment, furniture, wheelchair, etc.
7. Clothing
8. Seminars, workshops, training/education
9. Books, periodicals, films
10. Job recruiting, vacancy, availability
11. Classified section (entire section)
12. Unable to determine (please describe)________

15 — Physician (circle one)
0. no
1. male
2. female
3. both
4. indeterminate sex

16	Nurse/lab technician (circle one) 0. no 1. male 2. female 3. both 4. indeterminate sex
17	Patient (circle one) 0. no 1. male 2. female 3. both 4. indeterminate sex

Sampling

Finding one biased textbook does not necessarily indicate a serious problem with school books in the community; nor is an analysis of one speech sufficient to confirm a political trend. A larger number of school textbooks or speeches must be analyzed. A single edition may be affected by extraneous factors, such as the time of year or some unusual event that may temporarily obscure long-standing policies. The larger the amount of material analyzed, the more valid the conclusions will be. While the more-is-better rule applies in general, at some point the cost of increasing the amount of material will outweigh the benefits. If a monthly periodical is being examined over a 10-year period, there will be 120 issues available for study. For reasons of economy, the researcher may decide to reduce this number through sampling rather than use of all of them. Probably a sample of six issues per year would be sufficient. However, the sample must be representative. The investigator cannot simply use those articles or speeches that support his or her beliefs.

There are several different ways to obtain a representative sample of material. The researcher may decide to use every third issue of a magazine or every fifth article over a 2-year period. The important consideration is that the choice of material be representative rather than under the arbitrary control of the researcher.

Decisions must also be made about the unit of analysis. One can analyze sentences, paragraphs, pages, or entire articles. In the drug advertising study, some ads ran more than four pages. In counting the amount of drug advertising should one four-page ad be scored the same as four one-page ads? Is a large editorial cartoon on a topic equivalent to a small one? One approach is to score each item in two ways. First, count each article, advertisement, or cartoon as a separate unit regardless of size (e.g., the number of antipolitician cartoons on the editorial page regardless of their size). Second, tabulate them by the amount of space (column inches) they occupy.

Reliability

Scoring categories must be reliable. Two people doing a content analysis of the same article using a single list of categories should come up with similar results. Reliability is greatest when the scoring categories are clearly stated and do not overlap. Reliability cannot be assumed. Even something presumably as obvious as the sex of advertisement models can be ambiguous. Several of the medical ads showed models who were so blurred and indistinct that their gender was indeterminate. There were even questions about whether some ads were in color or black and white. Sepia tone ads were difficult to classify as to color.

The first step in checking reliability is to train two or more judges in the use of the scoring system. It is important that the training be leisurely and relaxed. The judges must be free to ask questions and express their doubts when they are uncertain. Even a good scoring system may seem unreliable if the judges are not properly trained. Training begins with a detailed explanation of the scoring system and then practice in scoring material under the watchful eyes of the investigator. The material used should be similar to that in the actual research project. When the judge has a problem, this can be discussed with the investigator. It may be that the categories are ambiguous and should be changed. At other times, the problem will be the judge's inexperience in using the system. If the judges still have problems after a few practice trials, then the classification system should probably be revised. It is wasteful to begin a lengthy content analysis if the categories are not reliable.

Limitations

The results of a content analysis are purely descriptive. They reveal *what* the content and structure are but not *why* they are that way; nor does the analysis reveal the impact of a communication upon the audience. This requires further investigation through observation, interviews, or experiments.

The researcher using content analysis is restricted to the materials available. Some topics are more likely than others to be covered in the media. Someone who is interested in media stereotypes of accountants or steamfitters, for example, would have to search a long time and would probably find little that was applicable.

Content analysis deals with communication materials, not necessarily with outside reality. An increase in media coverage of robberies and burglaries may reflect less of an actual crime wave than an editorial decision to highlight crime during a particular month. The investigator must be careful not to assume that trends in written or audiovisual materials automatically reflect changes in the topic being covered.

Summary

Content analysis is a technique for systematically describing written, spoken, or visual communication. The analysis can emphasize either the content of the material or its structure. The basis of content analysis is quantification.

The researcher uses material that is already available; there is no need for interaction with an actual subject. The method is well suited for cross-national comparisons and for examining trends over time. The material chosen for analysis must be selected in an unbiased way, and reliable categories must be used in the scoring.

Limitations: The results of a content analysis are descriptive rather than explanatory. Analysis is limited to available recorded materials and is subject to any biases in these materials.

References

Bilu, Y. (1989). The other as a nightmare: The Israeli–Arab encounter as reflected in children's dreams in Israel and the West Bank. *Political Psychology, 10,* 365–389.

Healey, J. S., & Kassarjian, H. H. (1983). Advertising substantiation and advertiser response: A content analysis of magazine advertisements. *Journal of Marketing, 47,* 107–117.

Kassarjian, H. H. & Kassarjian W. M. (1988). The impact of regulation on advertising: A content analysis. *Journal of Consumer Policy, 11,* 269–285.

Moore M. (1984). Sex and acknowledgments: A nonreactive study. *Sex Roles, 10,* 1021–1031.

Further Reading

Krippendorff, K. (1980). *Content analysis: An introduction to its methodology.* Beverly Hills, CA: Sage.

Merriam, J. E., & Makower, J. (1988). *Trend watching: How the media create trends and how to be the first to uncover them.* New York: AMACOM.

Weber, R. P. (1985). *Basic content analysis.* Newbury Park, CA: Sage.

12 Personal Documents and Archival Measures

Personal Documents

With all the new household appliances presently available, one would suppose that American women would spend far less time in housework than did their mothers and grandmothers in the 1920s. Not so, concluded researcher Joann Vanek (1974), who examined diaries kept by women as part of a research study sponsored by the U.S. Bureau of Home Economics. Participants in the study had been asked to keep records of their daily activities. It was found that women not holding jobs outside the home spent as much time doing housework as did women in the 1920s. Diaries did show changes in the nature of housework. Time spent in shopping and administering the home had increased. Less time was spent in preparing food and cleaning up after meals. Washing machines, dryers, and abundant hot water made laundry activities much easier. Nonetheless, the amount of time women spent doing laundry has actually increased over the past half century. Hochschild (1990) estimates that women today work approximately 15 hours more each week than do men in combined employment, housework, and child care. They spend one shift at the office or factory followed by a "second shift" at home.

These conclusions could not have come from the laboratory, but rather from the study of people in their homes. Up to this point, the major procedure recommended for studying natural behavior has been observation. Another possibility is to ask people to keep journals of their activities or to use diaries that people keep on their own. This method is advantageous when the behavior is private and difficult to observe without being intrusive.

Research Diaries

Researchers have asked people to keep diaries in many types of investigations. Social influences on eating were studied by paying 63 people to maintain week-long records of everything eaten, including the time of day, subjective ratings of hunger, and the number of people present. Meal size was positively related to the number of people present (de Castro & de Castro, 1989). To investigate how people adapt over time to an unfamiliar complex situation, psychologist Gerhard Kaminski (1987) asked two of his students to take unfamiliar jobs and keep a daily log of their experiences. Every evening for 2–3 months, each student dictated an account of the day's happenings into a tape recorder. The tapes were later typed for use in a detailed content analysis. Diaries were also used by a research team interested in knowing how much time undergraduates spent studying. This answer, too, could not come out of the laboratory. To find out how long students actually study, one must enter the world of dormitories, libraries, and cafeterias. Researchers considered a number of other techniques before deciding on journals kept by individual students. The lack of any specific location for reading discouraged the use of observation. Reviewing notes was found to occur at odd hours—in between cups of coffee in the cafeteria, lying on one's bed listening to the stereo, at the laundromat, and occasionally in the bathroom. There was no single place from which an observer could keep track of all the reading that was done. Interviews about the amount of study during the past week were also unsuccessful. Because of the varied nature of study and the fact that it was frequently combined with other activities, students were vague and hesitant in their answers. This interviewer received replies like, "Two or three hours, I guess," or "Maybe five hours, I don't know." Another approach considered but then discarded was to track students throughout the day to observe their study habits directly (i.e., behavioral mapping). This would have been extremely intrusive and time-consuming. The presence of an observer in a small dormitory room would have been intolerable to many occupants; nor would many students care for an observer accompanying them on a study date. The diary was considered a good compromise between accuracy and nonintrusiveness, especially since the diaries could be supplemented by observations in the library and interviews with students about their study habits.

Most research diaries are limited to specific activities. Few researchers ask people to keep open-ended journals describing everything that has happened. Such accounts would be extraodinarily tedious for both the diarist and the researcher. Most research diaries have time periods indicated directly on the record sheet. The investigation of housework used 5- and 15-

Personal documents. Because studying so often took place at odd hours in odd locations, researchers asked students to keep diaries showing how much and where they studied.

minute intervals, and the research on study habits used hourly intervals (see Box 12-1). Finding appropriate time units is very important. Information will be lost if the intervals are too long, but if the intervals are too short, the amount of paperwork increases and cooperation is likely to suffer. When the activity has a clearly defined beginning and end, logs can be kept in minutes (e.g., telephone calls took place between 9:18 and 9:21, 9:36 and 9:45) and so on. With activities that are loosely defined and subject to intrusion, records can specify the primary activity that occurred during the period.

Life Histories

The diaries mentioned thus far in this chapter were prepared at the researcher's request. Other sources of information on a topic are published diaries, journals, letters, and autobiographies. In some cases, people have published their own journals, but in other cases it is someone else—perhaps a relative, friend, or historian—who locates an interesting diary and obtains permission to get it published. A researcher, studying grief and bereavement, analyzed diaries of 56 people who experienced the loss of someone close to them. Quantitative and qualitative analyses were used to compare grief

BOX 12-1. Record Form for Study Sessions

Day__________ Date__________ Initials or Code Number__________

Time	*Amount of time (in minutes) spent in:* Reviewing notes	Reading text	Reading other materials	Preparing term report, project, etc.	Other (describe)	Length of study periods	Place where study occurred
8–9 A.M.							
9–10 A.M.							
.							
.							
.							
11–12 P.M.							
12–4 A.M.							
4–8 A.M.							

Comments:

responses in the nineteenth century with those of the twentieth century (Rosenblatt, 1983).

Letters written by famous individuals have also found their way into print—for example, Freud's letters to Wilhelm Fliess (Freud, 1957) and Van Gogh's letters to his brother Theo (Van Gogh, 1953). Originally written as private correspondence, they have become public documents available for journalistic, historical, and psychological studies. There are published autobiographies relevant to almost any conceivable topic. Occasionally a sufficient number of autobiographies is available to permit systematic analysis. Numerous scientists and inventors have written accounts of how they made their discoveries. Politicians frequently write autobiographies, as do film stars, singers, musicians, and artists. Over a hundred published autobiographies of prison inmates are available. These accounts cover all time periods and many nations, permitting comparison of prisons across centuries and national boundaries.

Relevant personal narratives are often found in unusual places. For example, interesting material on medical practice can be found in the autobiographical poems of poet/physicians. A good account of blindness is Eleanor Clark's *Eyes, etc.* (1977). Reading first-person narratives is easy and enjoyable. Most can be finished in a single evening.

A reference librarian can be extraordinarily helpful in locating pertinent autobiographies. It is surprising how many bibliographies (lists of books on a topic) and books about books are published each year. Someone has undoubtedly written a master's thesis on the literature of the steel mills of the nineteenth century or the building of the transcontinental railroad. We cannot emphasize too strongly how helpful a librarian can be. An entire profession exists whose avowed goal is to help you gain access to books. Do not hesitate to consult the owner of a used-book store, who is likely to be a book lover. Book dealers can initiate searches through trade newsletters. Purchasing secondhand books is not necessarily expensive as long as you avoid rare editions. For instances, two researchers collected autobiographies of former mental patients using the newsletters of book dealers at an average cost of $5 per volume. The key in locating published autobiographies is to use the services of available book people such as librarians, book dealers, archivists, and historical society members. When they do not know where materials can be found, they will often be able to supply the name of someone else to ask for advice.

The most common use of autobiographies in behavioral research is to illustrate findings that have been established through other methods. One social psychologist interested in sensory deprivation quoted the experiences of early arctic explorers. Researchers who study creativity will include statements by famous inventors. Reports on crime are likely to quote from the diaries of famous swindlers, burglars, and pickpockets. Care must be taken to ensure that the statements are not taken out of context or selected in such a way as to ignore contrary information. The researcher who quotes a prison inmate's description of food as "slop" should be prepared to document this description in other prison writing. This particular description holds up, since virtually every account of prison life describes the food as overcooked, tasteless, and nonnutritious. The most usable quotations have both internal and external validity; that is, similar statements occur elsewhere in the same diary and in other people's journals.

Fact or Fiction?

The line between autobiography and fiction is often blurred. Ken Kesey's novel *One Flew Over the Cuckoo's Nest* (1962) was based on his experiences as an attendent in a mental hospital. Joseph Wambaugh's *The New Centurions* (1970) was based in part on his work in the Los Angeles Police Department. Are such books fact or fiction? A little of both, probably. It is more hazardous to rely on fictionalized accounts than on autobiograpies. A novelist is freed from the constraints placed on a person who claims to tell a truthful tale.

A biography is one person's life as seen by another person. Its value as a research document depends on the skill, diligence, and accuracy of the writer. Subjects of biography tend to be even less representative of people in general than are subjects of autobiographies. Few biographers will spend time on an ordinary lawyer or politician. In seeking examples, ideas, or personal narratives for a research paper, our preference is for autobiographies over biographies and for avowedly factual to fictionalized accounts.

Another type of biographical study is the sociological life history obtained through intensive interviews. These personal narratives can add fascinating dimensions to behavioral research, and they provide balance to the biographical studies of more famous or powerful people. Oscar Lewis' accounts of family life and social relationships in Mexico are as compelling as a well-written novel. *The Children of Sanchez* vividly portrays through the words of the respondents, the lives, hopes, despairs, and dreams of the poor (Lewis, 1961). The richness of the accounts reflects a careful distillation of hundreds of hours of interview notes and tape recordings. Lillian Rubin (1976) explored life in working-class families in the San Francisco Bay Area. Robert Coles (1977) illuminated our understanding of minority children in his sociological life histories of *Eskimos, Chicanos, and Indians.* These accounts of personal history are subjective. Although they can be quantified by means of a content analysis, much is lost in the process. Retaining them in narrative form provides *qualitative* data—information not reducible to quantity or amount. It is often difficult to demonstrate the reliability and generalizability of such accounts. Despite this drawback, qualitative information complements quantitative data in increasing our understanding of behavior and social process.

Private Documents

Occasionally a researcher gains access to letters or documents not intended to be made public. Letters of resignation and written complaints fall into this category. A personnel manager may have a file of complaints. The personnel director of a restaurant may have exit interviews with departing employees. Such records may be valuable sources of information about personnel procedures or health and safety practices. Their use in social research was pioneered by W. I. Thomas, a Chicago sociolgist, who believed that such materials revealed people's views of themselves and of the world (Thomas & Znaniecki, 1958). For his comparison of rich and poor people in Chicago, Harvey Zorbaugh (1929) used as sources letters, personal diaries, and suicide notes. A more systematic study of suicide notes was made by Shneidman and Farberow (1957), who used them to understand the thought processes of people attempting to kill themselves.

Limitations

Published autobiographies are frequently biased. The authors' intentions in writing were to place themselves in a good light or to support some cause. Such people are generally atypical. They are more articulate and intelligent than others who have undergone the same experiences but have not written about them. Police officers who have written about police work are not typical officers. The same is true of the airline stewardess who wrote *Coffee, Tea, or Me* and the school teacher who wrote *Up the Down Staircase*.

Although personal documents can be analyzed in the same way as other written materials using content analysis as described in the preceding chapter, this should be done only with extreme caution. What does it mean that 18 out of 20 letters received by the City Parks Department are complaints? Statistical analysis of such letters is a poor guide to public attitudes. Bias can be assumed even when most of the letters and documents are favorable. Most of the people who write spontaneoulsy on a topic will feel strongly pro or con. Personal documents not intended for publication should be tabulated and analyzed only with extreme caution and with respect for the privacy of the people involved.

Archival Measures

Archives are public records and documents. They contain a wealth of information for the researcher. Over a century ago, the French sociologist Emile Durkheim (1951) examined records of suicides from all parts of Europe where such records were available. He found that suicide rates differed greatly among nations. Those nations that had the highest rates during one time period also tended to have the highest rates at other times. Durkheim's statistics were used to test various hypotheses. A common belief was that suicide increased with the temperature. Durkheim's data did not support this hypothesis; there was no straightforward relationship between temperature and suicide rate. What seemed most important was the cohesiveness of the society. When family ties, social relationships, and political ties were strong, the suicide rate was low. When the social fabric began to come apart, the rate rose.

Many statistics are readily available in government reports; other information requires a diligent search through newspaper files, crime statistics, biographical directories, and census records. Most of the other techniques discussed in this book require the researcher to go out and collect original data. For the archival researcher, the data are already available in almost limitless quantity. This creates a high risk of having more information than can be sorted into meaningful categories. The researcher is faced with the task of choosing from among statistics from different regions and time periods.

Most reference libraries have statistical data from public records, including census statistics. There are also yearbooks, sourcebooks, and factbooks with public statistics on a wide range of topics. For information on the United States, the *American Statistical Index* lists the publications of more than 400 central and regional government agencies, providing access to consumer price index reports, education and employment statistics, and so on. The *Statistical Reference Index* provides a topical index to information gathered by state governments, universities, and private associations. These two indexes are available in bound volumes and on-line computer systems.

Public records can be used as a check against interview and case study material. Anthropologists Sylvia Forman and Richard Mazess (1978) matched baptismal records with the stated ages of residents of Vilcabamba, Ecuador, where many residents claimed to live more than a century. Earlier reports stated that residents lived to age 130 and that a large proportion was over 65. However, Forman and Mazess found no one older than 96 and an expected proportion of senior citizens. "It's clear that a few individuals were specifically lying about their age, but most of them believed what they were saying," the researchers reported. "There is apparently a natural inclination of the elderly in Vilcabamba to exaggerate their ages for reasons of status or celebrity . . . and they have been getting encouragement from both tourists and the scientific community to do that." Available records provided a check on the residents' self-reports.

Underreporting and Bias

Today many researchers are skeptical about the value of published statistics on suicide. Many families will conceal a self-inflicted death regardless of the method used by the victim. Some automobile accidents, particularly those involving a single automobile, are believed to be suicide attempts, although they are rarely classified as such. Whether or not death from drug overdose is classified as suicide is often a matter of chance. When there is underreporting of some condition, official estimates of its occurrence will be too low.

Researchers have dealt with underreporting in various ways. One method is to try to estimate the extent of error and correct for it statistically. Like suicide, incidents of robbery are also significantly underreported. The *victim survey* has been used to correct robbery statistics. Instead of relying only on police records, researchers ask a representative sample of people in a district whether or not they have been robbed during the past year. If 15 out of 100 residents report a robbery while police statistics show a rate of only 5 per 100, there is an assumed underreporting factor of 66 percent. If this error rate is found in other victim surveys, the police crime statistics in similar districts can be corrected by 66 percent to yield a more accurate picture of the number of robberies.

Another approach is to seek out those cities and agencies with the most detailed and accurate social statistics. For example, public health researchers in the United States have often been frustrated in keeping track of schizophrenics. When these poeple move, all contact with public agencies in the first city is lost. To get around this problem, researchers on schizophrenia make use of records in Denmark and Switzerland, where social statistics are more accurate and comprehensive than those in the United States. Records of birth, death, family relationships, and hospital admissions going back hundreds of years can be found in district offices. While information obtained in Denmark can be generalized only with caution to the situation of schizophrenics in the United States, it can contribute to a multimethod research strategy. The Denmark data have different limitations than studies of mental patients in the United States, which tend to cover a large geographic area over a short time period.

Another way of correcting for bias is to supplement social statistics on large populations with detailed case studies and interviews with smaller samples. Public records are invaluable in a multimethod approach. What they lack in depth can be compensated for by their greater comprehensiveness and coverage over time. Unlike data collected for a research study, which tends to be a one-time activity, government records are collected repeatedly. Once an agency begins publishing statistics on some topic, it seems difficult for them to stop. This is advantageous to the researcher who wants to compare trends over time. However, the researcher must check to see that the methods of data collection have not changed. For example, annual listings of major crimes provided by the Federal Bureau of Investigation have undergone several significant revisions, making it difficult to directly compare crime figures from two time periods. Further analysis is necessary to correct for bias resulting from changes in the method of data collection.

Because public records come in so many forms, no attempt will be made to discuss methods of analysis. Using census data or other social statistics to study characteristics of groups is termed *aggregate data analysis.* Specific references for handling these data are listed at the end of the chapter. In general, aggregate data are more useful for making predictions about groups than about individuals. Some distortion and error may be assumed in all statistics, no matter how carefully they have been collected and assembled. The important question is whether the errors rule out the use of the records in a particular study. Some statistics, such as those for victimless crimes, have been proved so erroneous that they are virtually useless. Records for robbery and rape are clearly underestimated, but this bias can be corrected to some degree. Some records are useful for certain groups in society but not for others. Census figures are generally considered more accurate for the white middle-class sections of the community than for racial and ethnic minorities and the poor. Census figures reveal very little about illegal aliens, who are heavily concentrated in specific areas of the country. Many older people sink into poverty and disappear from public records.

Finding Unusual Sources

In seeking information, the researcher is not limited to the most obvious reports, such as crime statistics, voter registration lists, and census reports. Numerous private firms keep detailed records of their activities. In studying the community response to a crime wave, it would be interesting to examine the sales of deadbolt locks and burglar alarms. Since no central records for this exist, it would be the researcher's task to visit locksmiths in the community and compare sales for the same month over several years.

The number of telephone calls between two cities provides an index of social contact. Most long-distance calls from Montreal go to other French-speaking cities in Quebec, the heart of French Canada. Matching for size and distance and adjusting for cost, a call from Quebec is five times more likely to be destined for another Quebec city than for a city in the adjacent largely English-speaking province of Ontario (Mackay, 1958). Mail and visiting records were used to study the desocializing effects of a long stay in a mental hospital. A newly admitted patient was 45 times as likely to receive a letter and 20 times as likely to have a visitor during a given week than was the patient who had been in the hospital several years (Sommer, 1958). Letters and visits indicate family interest, which is considered a valuable asset to the patient's recovery.

Limitations

The quality of record keeping varies significantly from one agency or city to the next. The classification system used in different jurisdictions may not suit the researcher's purposes. A felony in one state may be a misdemeanor in another. Methods used in gathering and classifying information are likely to have changed over time. Many statistics, such as crime and illness rates, tend to be underreported. Other figures, such as unemployment and cost-of-living statistics, are occasionally manipulated for political purposes. Researchers tend to be justifiably suspicious of statistics compiled by individuals and agencies with little interest or stake in accurate records.

Summary

Personal documents provide a subjective view of events. They are especially useful for understanding private behavior. Asking people to keep diaries of specific behaviors provides material for content analysis. Autobiographies, letters, journals, and diaries can be used to illustrate findings that have been established through other methods. The line between biography and fiction is often blurred. Some books are a combination of both.

Limitations: Autobiographies are frequently biased accounts by unusual individuals. Letters received by public agencies also have limited generality because they may not reflect the views of the larger population.

Archives include public records, documents, and statistical reports. Public records are often useful as a check against interview and case study material.

References

Clark, E. (1977). *Eyes, etc.: A memoir.* New York: Pantheon.
Coles, R. (1977). *Eskimos, Chicanos, and Indians: Children of crisis* (Vol. 4). Boston: Little Brown.
de Castro, J. M., & de Castro, E. S. (1989). Spontaneous meal patterns of humans: Influence of the presence of other people. *American Journal of Clinical Nutrition, 50.* 237–247.
Durkheim, E. (1951). *Suicide.* Glencoe, IL: Free Press.
Forman, S., & Mazess, R. B. (1978, March 10). They either lie or forget. *Daily Democrat* (Davis, CA), p A2.
Freud, S. (1957). *The origins of psychoanalysis. Letters, drafts, and notes to Wilhelm Fliess, 1887–1902.* (M. Bonaparte, A. Freud, & E. Kris, Eds.) Garden City, NY: Doubleday.
Hochschild, A. (1990). *Gender and grace: Love, work, and parenting in a changing world.* Downers Grove, IL: Intervarsity Press.
Kaminski, G. (1987). Cognitive bases of situation processing and behavior setting participation. In G. R. Semin & B. Krahé (Eds.), *Issues in contemporary German social psychology* (pp. 218–240). Beverly Hill, CA: Sage.
Kesey, K. (1962). *One flew over the cuckoo's nest.* New York: Viking.
Lewis, O. (1961). *Children of Sanchez: Autobiography of a Mexican family.* New York: Random House.
Mackay, J. R. (1958). The interactance hypothesis and boundaries in Canada. *The Canadian Geographer, 3,* 1–8.
Rosenblatt, P. D. (1983). *Bitter, bitter tears: 19th-century diarists and 20th-century grief theories.* Minneapolis: Univerisity of Minnesota Press.
Rubin, L. B. (1976). *Worlds of pain: Life in a working class family.* New York: Basic Books.
Shneidman, E. S., & Farberow, N. L. (Eds.) (1957). *Clues to suicide.* New York: McGraw-Hill.
Sommer, R. (1958). Letter writing in a mental hospital. *American Journal of Psychiatry, 115,* 514–517.
Thomas, W. I., & Znaniecki, F. (1958). *The Polish peasant in Europe and America.* New York: Dover.
Vanek, J. (1974). Time spent in housework. *Scientific American, 231,* 116–120.
Van Gogh, V. (1953). *Verzamelde Brieven,* Amsterdam: Wereldbibliotheek.
Wambaugh, J. (1970). *The new centurions.* Boston: Little, Brown.
Zorbaugh, H. (1929). *The gold coast and the slum.* Chicago: University of Chicago Press.

Further Reading

Allport, G. W. (1942). *The use of personal documents in psychological science.* New York: Social Science Research Council.
Denzin, N. K. (1989). *Interpretive biography.* Beverly Hills, CA: Sage.
Gottschalk, L., Kluckhohn, C., & Angell, R. C. (1945). *The use of personal documents in history, anthropology, and sociology.* New York: Social Science Research Council.
Runyan, W. M. (1982). *Life histories and psychobiographies.* Oxford, U.K.: Oxford University Press.
Stewart, D. W. (1984). *Secondary research: Information, sources, and methods.* Beverly Hills, CA: Sage.
Wasserman, P., & Paskar, J. (Eds.) (1977). *Statistics sources* (3rd ed.). Detroit: Gale Research.
Webb, E. J., Campbell, D. T., Schwartz, R. D., Sechrest, L., & Grove, J. B. (1981). *Nonreactive measures in the social sciences* (2nd ed.). Boston: Houghton Mifflin.

13 Case Study

In the field of cultural psychiatry, *koro* is a condition of panic regarding the possibility of genital shrinkage, coupled with fear of impending death. The condition has been reported primarily in the southern coastal provinces of China, occasionally as epidemics affecting hundreds or thousands of people. In 1984–1985 there was a massive koro epidemic involving more than 2,000 people in Guangdong, China. Teams of researchers from the Guangdong Psychiatric Research Institute and the University of Hawaii interviewed 232 koro victims (Tseng et al., 1988). Among the 232 cases, 84 percent involved males and 16 percent involved females. Most were between ages 10 and 25 years, the majority being adolescents. Because they embraced the folk belief that the shrinkage of the genitals was caused by the female fox spirit and could be fatal, most were already anxious when they heard the news about the spread of koro. Before their own attacks, about three-quarters had actually seen others having panic attacks and had witnessed other people attempting to assist a victim. Their own episodes usually began during the night. After the onset of chills, the males experienced a sensation that their penises were shrinking. Thinking this to be a fatal sign, they became panic-stricken and pulled at their penises while simultaneously shouting for help. They experienced anxiety (100%), tremors (69%), palpitations (65%), shouted in fear (68%), and were preoccupied with thoughts of impending death (62%). The female victims usually complained that their nipples were retracting or their sexual organs were shrinking.

Upon witnessing the episode, family members and neighbors reacted decisively in the belief that the victims were in need of rescue. They would physically pull at the sexual organs fearing that the genitalia might retreat into the abdomen and cause death. The victims were fed red pepper jam, black pepper, or ginger juice.

In the majority of cases, the attack was brief and ended within 20–60 minutes. Only a minority (22%) had a second attack and very few had more than three attacks. After the panic subsided, most of the victims recovered completely. No one died of genital retraction, but there were several injuries due to mistreatment by others trying to be helpful.

Advantages of a Case Study

Unusual events like this call for innovative techniques. When there is no opportunity for a controlled experiment or before-and-after observations, the researcher may still want to undertake a careful investigation after the fact. A *case study* is an in-depth investigation of a single instance. It can involve a unit as small as an individual or as large as an entire community or region. It provides the opportunity to apply a multimethod approach to a unique event or setting. Unlike other methods that tend to carve up a whole situation, community, or life into smaller parts, the case study tends to maintain the integrity of the whole with its myriad of interrelationships.

The case study has a long and honorable history in clinical medicine. What is lacking in breadth and generalizability may be compensated for by greater depth. When a physician comes across unusual symptoms, it is important to examine the patient's background and experience in great detail. Has any member of the patient's family shown anything like this? Has the patient been exposed to toxic substances? Any recent foreign travel? If there appears to be a connection between the symptoms and some specific factor, a report may be published in a medical journal even though it is based on a single set of observations. The author hopes that others have come across patients with similar symptoms and can supply confirming or negative instances. When several case reports link the symptoms to a specific cause, then it is time to undertake a survey or experimental study. The first case report can begin a chain of events that produces an important discovery.

Another use of a case study is to test a theory. While a single exception is not sufficient for discrediting a theory, a supportive finding increases confidence in a theory's predictive power. For example, many of the principles of children's intellectual development can be illustrated with individual cases. A case study of the rise, growth, and decline of a corporation may lend support to one of several competing theories of organizational development.

Choosing a Topic

There are no rules for selecting a topic worthy of detailed investigation. Most researchers choose things that are unusual and newsworthy, such as natural disasters, serious illness, riots, fads, and fashions. The method has also been used to study the acceptance or rejection of innovations. This has created the false impression that a case study can be used only with unusual events. There is no theoretical reason why the method cannot be applied to ordinary people doing ordinary things. The practical objection, of course, is that it is likely to make dull reading and no one will publish it. However, studies that may seem ordinary at the beginning can be of considerable scientific value. Piaget's (1952) pioneering studies of cognitive development began with his in-depth observations of his own children.

In the study of an unusual event, case studies of other individuals or communities who have experienced it can be collected. The desire for multiple case studies from different disasters has encouraged international collaboration. Disaster researchers at Georgia Institute of Technology have an ongoing collaboration with their counterparts at Kyoto University in Japan (Abe, 1982). Researchers in Norway and Sweden have been studying railway engineers involved in accidents causing fatalities (Leymann, 1989). The small number of individuals and the unique circumstances connected with each accident have some of the qualities of a case study. The accumulation of individual accounts and the availability of information about the employees prior to the accident and of their fellow employees not involved in accidents also give the studies some aspect of a quasi-experiment or experiment-in-nature.

A case study has a temporal dimension—it shows the changes that occur over time. It also covers a wide net of informants and situations. This will probably require more time and effort on the part of a researcher than a comparison of two groups tested once (e.g., comparing the questionnaire responses of 20 individuals who had been in a natural disaster with responses from a control group, especially if testing can be done in a group session). The two approaches, the case study and the group comparison, produce different types of information, but ideally will be complementary.

The case study is widely used in anthropology to study cultures, in sociology to study communities, and in clinical psychology to study individuals. Educational researchers employ the technique to study innovations. Unlike before-and-after measurement, which examines changes on an objective test at two points in time, the case study focuses on the processes of change with attention to the role that individuals play in promoting or hindering a new program. For example, a case study may reveal that the innovation failed because it was imposed arbitrarily on key individuals who refused to support it, not because of its form or content. The richness and breadth of the material in a case study can help to explain the findings obtained using other methods.

Case studies provide illustrative examples within larger investigations using multiple methods. A case study can humanize a more statistically based piece of research and increase reader interest by connecting statistical findings to real-life examples. The koro study was not only about victims in the abstract, but about Mr. Gwee, an unmarried 28-year-old office worker who suffered a koro attack while preparing for an examination. One evening he heard sounds of a gong informing the populace of a panic in a nearby neighborhood. He suddenly became anxious and experienced his own koro attack. The researchers followed his life for several years afterward. He later married and advanced in his career.

The researcher's responsibilities do not end with collecting information. There remains the task of synthesis and analysis—putting together the information from various sources into a coherent whole. The case study generally (but not always) ends with some kind of synthesis or explanation. For each conclusion there needs to be evidence presented in the case study.

Obtaining Cooperation

The researcher's first task in a case study is to obtain the cooperation of the people involved. This is not as easy as it sounds. People who have gone through a traumatic experience are struggling to survive. A recent tragedy may be so intense that they do not want to talk about it. The researcher will not be able to offer people fame, fortune, or anything tangible. Probably the investigator will offer the very opposite—a firm guarantee of anonymity. The researcher promises to conceal the respondent's identity. The opportunity to help science may induce some people to cooperate even without tangible remuneration. However, the deciding factor is usually not financial gain or an abstract goal, but is instead a personal relationship with the researcher.

You will gain or lose people's cooperation by your manner of approach and interview style. An anthropologist interviewing natives in the Canadian Arctic had a difficult time explaining his purposes. The natives could not comprehend why he wanted to know the intimate details of their lives. He wasn't getting any cooperation until he admitted that his employer told him to get the information. *That* they could understand. Compliance with the boss's orders required no further explanation. The anthropologist was perceived as another person doing his job. In your own case study, you will find that people open up because they like you, trust you, and want to help you. This is particularly true if the study continues over a long period and involves personal revelations. Of all the research methods discussed so far, the case study places the most emphasis upon the researcher's style, approach, and personality.

With poor or needy individuals, money is helpful in obtaining cooperation. It conveys the researcher's seriousness of intent and the value placed on the information. Even with an affluent respondent, a token payment can be helpful. There is widespread belief in this society that something obtained for nothing is not appreciated. However, there are situations in which money cannot be used. For example, one could not offer $100 to the mayor or city planner to discuss zoning policies. Offering payment would be insulting, unethical, and perhaps illegal.

Avoid overselling your project in order to gain people's cooperation. Listen to their objections to your study before you begin. Encourage them to express any doubts they might have, such as fears for their privacy, the possibility of bad publicity for the community, and so on. It may be easier to work out ways to deal with these objections at the outset than in the midst of gathering data.

Make sure that you contact the key individuals involved with the issue you are studying. At the close of an interview, ask people specifically for the names of others who should be contacted. There is nothing more embarrassing after you have finished your study and left the area to find out that you neglected to interview important sources.

Cross-Verification

A description from a single observer should be regarded as tentative. When three or four people independently provide similar accounts, it seems more substantial. Cross-checking the accounts of independent observers is one means of assessing reliability in a case study. For example, a tornado victim may tell the researchers how he tried to help other people after the twister struck. Other accounts may indicate that the first individual was dazed and groggy and had to be led to safety. Do not conclude that the first respondent is an unreliable liar. Distortion becomes a topic of investigation. How consistent are people's accounts of their own actions during and after a calamity? We are led directly to a study of how people put unwelcome thoughts out of their minds.

A researcher should not challenge the respondent even when an account seems biased or incorrect. You are not the district attorney conducting a cross-examination. The goal is to obtain the respondent's point of view. Verification comes afterward. You must convey the impression of being a good listener. The respondents' knowledge that you have access to neighbors, newspaper reports, and public records will keep them from straying too far from the truth. Deliberate collusion among townspeople in concocting a false story is a possibility, but it does not occur often.

Differences between two accounts may reflect the way each person saw the situation. The person who spent several days in an evacuation camp after the tornado may have a different view of events than the person who stayed in town removing debris. Each of the views is accurate but is based on a limited range of experience. Inconsistencies in people's narratives can provide valuable leads. The researcher is a bit like the psychoanalyst who pays special attention to distortions and omissions. Asking for further elaboration is a better way of internally checking a story than challenging people as to their truthfulness or objectivity. Cross-verification is also possible through a multimethod approach. Perhaps better than all other techniques, the case study lends itself to the use of multiple sources and techniques for gathering information. Box 13-1 summarizes the skills needed by a case study researcher.

Observation

Case studies may also benefit from the systematic application of observational procedures. The case study of a child may include observations of behavior on the playground, in the classroom, or in other settings. For conducting a case study of an organization, participant observation (see Chapter 4) may be useful. Although there can be problems in being objective (i.e., avoiding observer bias), some researchers study organizations to which they belong.

Trace Measures and Physical Artifacts

A case study of a tornado would necessarily include descriptions of damage and dislocations arising from the event. Analysis of graffiti would be useful in a case study of teenage gangs. Childrens' drawings would be appropriate for analysis in a study of developmental processes. Case studies of some mental patients have included examples of their artwork. The cat paintings of Louis Wain, a famous British artist, showed his progression into severe mental disorder. The early cats were happy and socially active while the last ones exploded into color and energy, losing almost all feline features (Parkin, 1983).

BOX 13-1. Requirements for a Case Study Researcher*

The researcher must

1. Be able to ask good questions.
2. Be a good listener.
3. Be adaptive and flexible, recognizing unexpected situations as opportunities to gain more knowledge.
4. Have a firm grasp of the issues being studied, for example, planning, politics, disease, etc.
5. Be unbiased by preconceived notions.

*Yin, 1989

Public and Private Records

Because a case report is likely to involve a newsworthy event, public records are valuable sources of information. The researchers who studied the koro epidemic found hospital records that documented earlier cases of the condition. In other types of case studies, the most relevant records are census statistics, school district records, crime statistics, or the multitude of other available public documents. For supplementing interview or observational material, consider the archival sources described in Chapter 12.

With a little searching, you may gain access to private records, such as letters or diaries. See Chapter 12 on the use of personal documents in research. These can be very useful in shedding light on life in the community before the researcher came upon the scene.

Limitations

Generalization from a case study is necessarily limited. Often an event is selected because it is atypical. No matter how many people were interviewed in Guangdong, China, and no matter how much time was spent in the area by the research team, this was only a single koro epidemic. It is appropriate to draw conclusions from the data (e.g., the panic was brief in the vast majority of cases), but the findings cannot be generalized to other koro outbreaks without further study.

Because so much depends on the researcher's personality and approach, a case study is difficult to repeat. The situation of two researchers conducting *independent* case studies of the same event virtually never occurs. For some topics, generalizability (external validity) can be increased by multiple case studies. The earlier example of railway engineer accidents is a multiple case study (Leymann, 1989).

In case studies that take place after the fact, the researcher must depend upon people's recollections of events. After a crisis, memories are likely to be selective and distorted. With dramatic events, behavioral effects may continue for years, and it will be difficult to determine when the study should end.

Summary

A case study is an in-depth investigation of a single instance. This method emphasizes the individuality and uniqueness of the participants and the setting. Reliability is obtained through cross-verification of people's accounts. Multiple methods, including observation, the analysis of physical traces, and public and private records are useful in supplementing interview data.

Limitations: Generalization from a case study is necessarily limited. The event studied is likely to be atypical. People's recollections of past events are likely to be selective and distorted.

References

Abe, K. (1982). *Introduction to disaster psychology.* Tokyo: Science Publishing.

Leymann, H. (1989). *Psychological effects of fatal accidents on the track—the reaction of the railway engineer* (abstract of current research, Division of Social and Organizational Psychology). Solna, Sweden: National Institute of Occupational Health.

Parkin, M. (1983). *Louis Wain's Edwardian Cats.* London: Thames & Hudson.

Piaget, J. (1952). *The origin of intelligence in children.* New York: International Universities Press.

Tseng, W.-S., Mo, K.-M., Li, J.-H., Ou, L.-W., Chen, G.-Q., & Jiang, D.-W. (1988). A sociocultural study of koro epidemics in Guangdong, China. *American Journal of Psychiatry, 145,* 1528–1543.

Yin, R. K. (1989). *Case study research: Design and methods* (rev. ed.). Newbury Park, CA: Sage.

Further Reading

Habenstein, R. W. (Ed.) (1970). *Pathways to data: Field methods for studying ongoing social organizations.* Chicago: Aldine.

Riley, M. W. (1963). *Sociological research: A case approach.* New York: Harcourt, Brace, & World.

Withers, C. (James West, pseud.) (1945). *Plainville, U.S.A.* New York: Columbia University Press.

Yin, R. K. (1989). *Case study research: Design and methods* (rev. ed.). Newbury Park, CA: Sage.

14 Apparatus

Instructions to Subjects in a Perception Experiment

> I am going to show you some figures in the viewer. They will be the same figures that you saw in the first part of the experiment, either the whole figures or parts of them. Exposure times will be very brief, so be alert. I will give you notice every time that a figure is going to be shown. Please write down the name of the figure on the score sheet in front of you. Feel free to guess if you aren't sure which figure it is. If you don't recognize it at all, please describe in your own words what you saw in the viewer. Are there any questions? Then, we'll begin the series.

These were the instructions read to participants in a perception experiment. The subjects were secretarial students in a business school who were paid small fees for their participation. From the subjects' standpoint, the room would probably be described something like this: "Small, dimly lit, rather empty except for a table, two chairs, and a big wooden box that looks as if you are supposed to look into it. A few gadgets on the floor, but I can't make out what they are." The big wooden box was a tachistoscope, an apparatus for presenting visual material in very brief exposures. The viewer portion stood on the left side of the table; the timing mechanism and exposure switch were housed in a metal cabinet on the floor out of the subject's view. The viewer portion had an opening, with a forehead rest, into which the subject looked. There were two visual fields that could be presented to the subject. One was a square field of bright white light. When the experimenter pressed a level on the electronic timer, a picture would appear in the viewer for 200 milliseconds. The picture was seen at the same distance and intensity as the previous white light. Several features of this tachistoscope made it useful for the particular study:

1. The bright white field served two purposes. It provided orientation for the location at which the picture appeared, and it prevented afterimages following exposure of the picture.
2. The use of "neon daylight" tubes presented the colors of the pictures accurately and without distortion.
3. The electronic timing circuit used no relays and therefore was silent in operation.
4. Electronic timing is highly reliable (± 1 millisecond) if it is checked periodically.

Note to the reader: You need not be concerned about the technical aspects of the tachistoscope or the particular experiment. This example was used to show some of the considerations involved in selecting the right piece of apparatus for an experiment.

Borrow or Build Your Own

It may not be necessary for you to buy your own apparatus, particularly if you are on a limited budget. It is surprising how much good apparatus exists unused in the storerooms of research and teaching institutions. A piece of expensive equipment bought for a single experiment will gather dust afterward. It may also be possible to use someone else's apparatus on a time-sharing basis. Plates for measuring color blindness cost as much as $50, but they are widely available in the offices of optometrists and ophthalmologists and in personnel departments in industry. Searching for apparatus is good training for a researcher. If you decide to continue in behavioral research as a career, you are likely to be searching for apparatus, supplies, and funds throughout your professional life.

Your needs may be so specific that nothing suitable is available on the commercial market. Then you will have to build your own equipment. The tachistoscope described earlier was built by a research assistant in a psychology department workshop. Access to a good workshop is necessary if you plan to construct your own equipment. Most research institutions and psychology departments have a workshop supervised by a knowledgeable technician. With permission, you may be able to construct your apparatus in the workshop, or at least obtain technical advice. A shop technician is also a good source of information on what equipment is available and on its condition.

Apparatus Supply Firms

Most of the research described in this book requires no more equipment than a typewriter, a copy machine, and access to a calculator or computer.

Occasionally, further materials will be necessary. Laboratory research almost always involves special equipment. This may either be built by the experimenter for one-time use or it may be a standard machine that is available from a commercial supplier. Physiological research may use brain wave recordings from an EEG machine. Studies of group activities may use recorders or special-purpose machines for recording interactions.

Apparatus comes in many different models and price ranges. One can purchase a simple pencil maze, useful for testing coordination or learning problems, for less than $40, or a sophisticated electronic maze for 20 times that amount. The latter version provides the experimenter with a detailed record of the subject's every move. Often the simplest hardware is the best, but there are times when a more complex machine can yield results otherwise unavailable.

Many different kinds of apparatus for behavioral research are available commercially. Manufacturers of behavioral equipment publish catalogs listing items in stock or that can be custom made. The catalog of one major apparatus supplier includes such varied items as hand steadiness tests, plates for assessing color blindness, optical illusions, eye movement recorders ($4000 and up!), and biofeedback apparatus.

There are both major and minor suppliers of apparatus for behavioral research. Two of the larger suppliers that will send catalogs free upon request are the Stoelting Company (1350 South Costner Avenue, Chicago, Illinois 60623) and the Lafayette Instrument Company (P.O. Box 5729, Sagamore Parkway North, Lafayette, Indiana 47902). Many smaller firms specialize in custom-ordered apparatus and will quote prices on request. To locate an apparatus supply firm nearby, check with people doing behavoral research in your locality. These can include psychologists working for industry or at a university. If the opportunity arises, you can also visit the equipment display booths at the annual conventions of your regional psychological association. Equipment suppliers demonstrate their newest apparatus and have their current price lists available at these conventions.

Mechanical Failure

Any machine will eventually break down or wear out. Over a period of time, mechanical malfunction is the rule, not the exception. Jam-proof slide trays will jam, automatic cameras will give the wrong light readings, and computers will go down. This does not mean that human beings are more accurate than machines. No machine has been known to deliberately deceive its owner. Rather, it means that the possibility of error must be considered at every stage of the research, from presenting materials to subjects in the laboratory to data analysis. A book could be written about all the things that can go wrong with a slide projector. It is wise to carry spare parts, such as an extension cord and an adapter to convert three-pronged to two-pronged electrical plugs. Have a screwdriver or, even better, a Swiss Army knife for making repairs on the spot. It is not an evil spirit that breaks machines at the worst possible moment—for example, in the middle of a crucial meeting or during an experiment—but rather the pressures on the researcher, which cause minor but significant technical operations to be omitted, such as pressing the record button or loading the camera with film. Most researchers have come to realize that dependence on hardware can be a mixed blessing.

The Camera and the Recorder

There are three major uses for cameras and recorders in behavioral research:

1. *To present materials to respondents.* There may be some value in putting audio instructions on tape so that all the particpants will hear the iden-

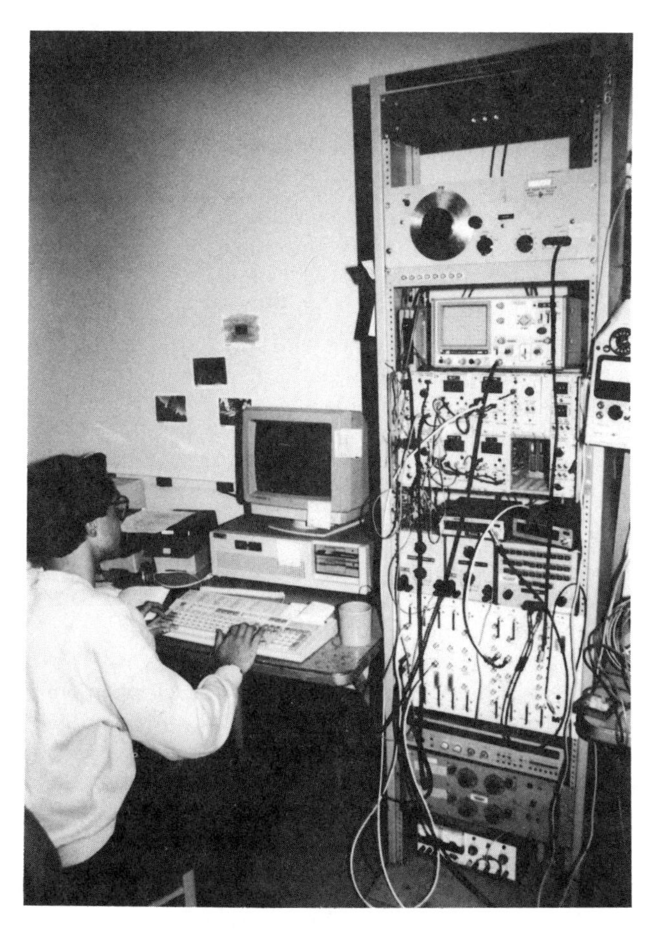

Equipment designed to measure neural activity from the auditory nerve. Not every experiment requires apparatus so elaborate.

tical words. The same is true of visual stimuli projected from slides or shown on videotape. The use of cassette recorders and slide projectors allows the researcher to standardize the method of delivery and presentation to a degree impossible any other way. Even when an interviewer deliberately tries to approach each respondent in the same manner, there are likely to be subtle differences from one time to the next depending on the interviewer's mood, the location, or even the time of day. Putting the introduction on tapes or slides minimizes these sources of error.

2. *To obtain data.* When researchers use cameras and recorders to obtain data, they can expect to spend a great deal of time typing transcripts of taped sessions and analyzing records of interaction, body posture, and so on. For this reason, obtaining data is *not* the most common use of cameras and recorders in behavioral research. The most common uses, in fact, are presenting materials to the respondent and illustrating the results obtained through other research methods. Only a small percentage of behavioral studies uses photographs and recordings as primary data because of the cost and effort involved in transcribing materials. While it is relatively easy to record a group conversation, obtaining a typed transcript is difficult and time consuming even under the best of circumstances. Unless you have a special audio tape recorder with a foot-reverse pedal (and very few tape recorders have this), the typist will have to reverse the tape by hand whenever a word or phrase is unclear. Since most conversations are full of half-spoken and half-heard phrases, transcribing an audiotape accurately will involve as many as 8 hours per 1-hour tape, with many words or phrases omitted entirely or incorrectly transcribed. Unless there was an observer with the tape machine keeping written notes of who was speaking, the typist will have difficulty identifying people's voices, particularly when only brief statements or interjections are made. This problem is reduced in the case of videotape, but the process of transcription is equally cumbersome. The presence of a recorder may inhibit some participants and induce them to avoid sensitive topics. The machine will run out of tape, the cassette will have to be changed or turned over, and this will disrupt the event. However, the most serious disadvantages of audio and video recording are the time and expense of transcription. Unless you yourself have tried to do this, you can't realize how boring, time-consuming, and frustrating transcription is.

In his sociological investigation of auto repair work, Douglas Harper (1987) creatively combined photographs with interviews in what he called a "photo elicitation" technique. Within an interview format, the subject of the pictures (in this case, the garage owner) and the interviewer discuss the pictures. Roles are reversed as the subject of the photograph becomes the teacher, and the interview takes the researcher in new directions, as the subject points out what is missing in the photographs and what should be included in future pictures.

Few behavioral researchers have used still or movie cameras as primary data sources. Transcription problems are not as severe as with tape recordings, since photographs can be developed and printed commercially. Large prints, which may be necessary for analyzing interaction, will run up the costs. Another problem with the still or movie camera is that, like the recorder, it is very intrusive. Many people do not want to be photographed. Zoom and telephoto lenses make the researcher's life easier. Movie cameras and videotape capture ongoing actions but the scoring problems that arise during transcription are equal to or greater than those involved in analyzing audiotapes.

Time-lapse photography can be more economical and efficient than videotape. The experimenter can set the camera to take a single frame every 30 minutes, 60 minutes, and so on. Whyte (1980) made extensive use of this in studying city life. Positioning the camera above the street in a second or third story window was a very economical way to study behavior at ground level. Several cameras could take in different areas simultaneously and produce archival records for detailed analysis. Whyte is particularly fond of small, light, Super 8 cameras that can cover any desired angle of view from telephoto to wide angle, take pictures at intervals from half-seconds to ten minutes, automatically adjust the light changes, and run unattended for up to 48 hours. The use of a fast lens and high-speed film allowed night photography. A digital clock, placed to appear at the bottom of the picture, provided a detailed record of the time when the frame was taken. This was very important when the film was analyzed and the researcher wanted to find specific people or situations.

The most difficult aspect of using time-lapse photography in research is the analysis of a large number of pictures. Whyte put it succinctly—time-lapse photography does not save time; it stores time. During the analysis, he found it most useful to run through the frames quickly, back and forth, to pick up major trends. This compresses time and helps the observer to identify patterns that might otherwise be missed. Once these patterns are noted, the major effort will be devoted to detailed analysis, sometimes frame-by-frame.

3. *As illustrations.* The camera is invaluable for illustrating results. When you are trying to describe what a town looked like after the tornado hit or the lack of privacy in dormitory rooms, slides and photographs will be helpful. Douglas Harper's (1987) participant observations in a small rural repair shop were greatly enriched by photographs showing the organization of space inside the shop, the exterior yard, and the junked cars waiting to be cannibalized for future projects. Sequences of photographs were used to show how the shop owner organized the repair process, trying to see things as he saw them, to share the mechanic's perspective. The book contains over a hundred photographs coordinated with descriptions and interview data.

Audio recordings are less useful than photographs for presenting results to an audience. Unless an audio recording is professionally made and edited, there are likely to be fuzzy and unintelligible spots that are confusing and frustrating to the audience. Editing requires expensive equipment, trained personnel, and much time and effort. Some audiences become restless and bored hearing an audiotape without seeing anything on the screen. Audiotapes are not as useful as slides for presenting the results of behavioral studies.

When using a camera in field work, you must plan for adverse conditions. Picture taking is generally forbidden in bars, gambling establishments, prisons, and hospitals. Adverse environmental conditions can include rain, power failure, and illumination too dim for most natural light photography. There are three solutions to the low-light problem. The first is the use of an inexpensive strobe attachment, whose chief disadvantages are distortions in color and light reflection as well as intrusiveness. It is difficult to remain in the background when you are using a strobe. Another possibility is to use high speed films and "push processing." The speed of the film is written as the ASA number on the side of the film. The higher the number, the more useful the film will be under low light conditions. If necessary the film speed can be "pushed" even higher through special processing. You should discuss this with a camera store owner before you take pictures to learn the maximum "pushing" that can be handled locally. The third method for handling weak illumination is to use time exposures. This requires a tripod along with a shutter release. Good pictures are possible in a dimly lit area if exposure time is sufficiently long. The chief disadvantage is that any movement will blur a portion of the picture.

Wind, rain, and snow create problems for a field photographer but they are all solvable with the right equipment and advice. Cameras made for underwater photography can be used in the rain. Cheap protection can be obtained by wrapping the camera in a transparent plastic bag. This will produce some distortion, perhaps not unpleasant from an aesthetic visual standpoint. There are also methods for making sound recordings under adverse weather conditions. Advice can be obtained from local radio personnel or audiovisual specialists.

Helpful Hints in Using a Still Camera in Behavioral Research

A supersophisticated and expensive camera is unnecessary, but on the other hand, an Instamatic is unsuitable. You'll probably need a single lens reflex camera that will cost about $200 for the basic camera and lens.

Inexpensive lenses are adequate for most behavioral studies. The need is for functional pictures rather than works of art. Besides the standard 50 mm lens that comes with the camera, you will probably want a zoom lens for special circumstances. The most useful zoom is probably one that covers a range from modest wide angle (35 mm) to modest telephoto (80 mm). This can be used with a teleconverter that doubles the focal length of a lens. Teleconverters are available in any good camera store and list between $20 and $35, depending on lens quality. You can also obtain a macro zoom lens that allows you to photograph close up and to copy printed matter. The cost of the same lens varies greatly from one camera outlet to another. So, be sure to shop around before you buy, and check the prices listed by advertisers in reputable photography magazines. *Consumer Reports,* which is available in most libraries, provides detailed evaluations of photographic equipment.

There are ethical and legal considerations in photographing people as part of a research study. Since laws vary from place to place, it is worthwhile to check with a local photographer before undertaking a research study involving camera work. A professional photographer can also supply valuable assistance in dealing with unusual lighting or haze conditions.

While tiny cameras hidden in matchboxes are important in spy movies, you probably won't need them in your study. Researchers who use a camera professionally tend to rely on telephoto lenses operated at a distance rather than spy cameras up close.

Helpful Hints in Using a Recorder in Behavioral Research

Audio Recorder

An inexpensive tape recorder can do most of the things an expensive one can do. You don't need tremendous fidelity for most purposes. The omnidirectional (all-directional) microphone that comes with most tape recorders is satisfactory for recording the normal range of voice frequencies. You will probably want to obtain an inexpensive stand to avoid holding or laying the microphone on a table where it will be frequently jarred. *Consumer Reports* is also a good source of information about audio equipment.

Try to obtain a recorder that can be battery-operated as well as plugged into an outlet. This will allow greater flexibility in your operations. Be sure to carry an extra pack of batteries and extra cassettes.

If you are going to have the tape transcribed, check on the availability of a machine that has a *foot reverse* pedal. This is an expensive built-in feature that most tape recorders lack. Note that we are talking about a foot *reverse* pedal and not simply a foot pedal.

If you want to play tapes to an audience, you will probably need a good player with at least two speakers. The separate speakers will enable you to aim the sound to different parts of the room. A good amplification system makes the difference between a clear representation and a confused, bored, and irritated audience.

As with spy cameras, there are very few situations in which we would recommend the use of hidden microphones. There are serious ethical problems in using them.

An audio specialist can help you select the right equipment for a special recording situation. Some microphones, for example, completely block out background noise; others record background and foreground equally. Know what is available and choose accordingly.

Videocamera

Videocameras are becoming increasingly common in research laboratories, especially when there is field work (outside study) to be done. Also, facilities for showing videotapes are becoming commonplace in classrooms and conference centers. When used to gather data, a videocamera is particularly useful for recording complex interactions that are difficult to observe and analyze at the time they occur. Behavior can be recorded at one time and analyzed later when the observers are available. Material can be played many times so that fine detail and complex interactions can be accurately noted. Early models were large, heavy, and bulky, and primarily useful for studio recording. Newer models are lighter, more portable, and more flexible for research use. If the camera is to be carried into the field, check out the weight with the battery pack. See that it comes with a shoulder strap or harness to aid in carrying it.

The technical aspects of videotape recording have also improved. Now this is a very convenient method for collecting information. The researcher zips in a cassette and the machine takes over. Cameras may also be wall mounted or used with tripods for stationary recording. Videocameras differ in special features and some are much easier to use in the field than others. Special lenses are available for the cameras, including wide-angle lenses useful for indoor recording, telephoto lenses for long-distance pictures, and macro lenses for close-up work. Some machines have built-in microphones and head sets for continuously monitoring audio quality. Other models allow for remote operation both in recording and playback. Some cameras allow for the insertion of titles, date, and elapsed time, and combine both recording and playback systems.

It is more common in behavioral research to code videotape data (as in a content analysis) than to transcribe it literally, which would be very arduous and time-consuming (i.e., describing every word and movement). See Chapters 8 and 11 for discussions of coding procedures. Coding videotapes is more complicated than dealing with most other types of material because they include both verbal and visual material. One potential disadvantage of videotape recording is the tremendous amount of information collected. This can encourage the researcher to postpone dealing with the tapes. The results will be a huge library of uncoded and unanalyzed videotapes.

Taking pictures of everything going on can be very wasteful and time consuming for data analysis. A better approach is to have some idea in advance of what you want to record. Viewing angle is very important, especially if there is to be a systematic analysis of the data that requires consistency in what is being recorded from one time to the next.

Videotaped interviews require planning and a warm-up period. Don't go into a setting cold and expect to interview people there. Check out the location in advance, particularly the lighting and sound environments. Is there too much background noise from a ventilating system or street traffic? Such considerations should be checked out prior to the formal taping session. No matter how well you know the person and the place, the presence of a camera can change the quality of interaction and a warm-up period will probably be needed. Arrange beforehand to have some friendly and interesting questions or comments to offer before the filming, or in the early stages. The important thing is to get the respondent's attention off the camera and on to the interaction with you. A shoulder- or hand-held camera is both awkward and intimidating during an interview. It is preferable to keep the camera off to the side while talking to another person.

Videocassette Recorder (VCR)

A VCR can be used to record programs made earlier or for presenting videotape selections to an audience. These capabilities have a number of applications in behavioral research. Programs recorded on network television can be subjected to a content analysis (e.g., the length and type of commercials, violence on prime-time shows, etc.). A VCR is also used for playing back tapes made with a videocamera. The various types of VCR machines are not necessarily compatible with one another or with home television sets. Professional models cost more than consumer models and produce a higher quality picture. Tapes for professional models are usually more expensive than those for consumer models. The choice of machine for research depends on the needed quality level and compatibility between recorder and playback machine. A videotape made on a Beta-type machine must be shown on a Beta-type VCR.

Some machines allow the possibility of playing tapes in slow motion. This may be a useful feature in interaction analysis. A "search mode" enables the researcher to locate needed items on a tape visually. For any type of content analysis, a PAUSE or FREEZE FRAME control will allow for the extended analysis of single frames. In some machines the frozen frame may be too fuzzy or unstable for serious analysis. For on-air recording, a counter will be helpful for making notes of significant features for later analysis. Video printers are available for making hard copy from video images. Slides and movie films can be transferred to videotape for presenting research results to an audience.

Summary

Apparatus supply firms manufacture equipment for behavioral research. Sometimes researchers must build their own equipment.

Still cameras, audio recorders, videocameras, and videotape cassettes are often useful in behavioral research. The camera and audiotape are used more often for presenting material to respondents and for illustrating talks presented to audiences than for obtaining primary data. Videocameras and videotape cassettes are often used in field study for recording complex interactions for content analysis and other purposes. Limitations: The cost of transcribing audiotapes and videotapes makes these methods cumbersome and time-consuming for data collection and analysis.

Reference

Harper, D. (1987). *Working knowledge.* Chicago: University of Chicago Press.

Whyte, W. H. (1980). *The social life of small urban spaces.* Washington, DC: The Conservation Foundation.

15 Standardized Tests and Inventories

Not surprisingly, many students are anxious before taking exams. To investigate this phenomenon, a researcher administered a standardized test for state anxiety (which is connected with a mood or situtation) to a research methods class early in the semester. The inventory used, the State-Trait Anxiety Inventory, consisted of 20 statements, such as "I feel calm" or "I feel upset" to be rated by the student in terms of how well the term applied: not at all, somewhat, moderately so, and very much so (Spielberger, Gorsuch, & Lushene, 1970). This inventory yields an overall state (situational) anxiety score.

A few weeks later, on the day of the first exam, the inventory was given again. A comparison score from the two administrations showed a significant increase in state anxiety on the day of the exam. However, there was no relationship between the amount a student's anxiety had increased and his or her actual exam performance. That is, students whose anxiety levels soared did neither better nor worse on the average than students whose anxiety levels changed very little. Also, there was no relationship between anxiety on the day of the exam and how well the student performed on the exam.

What Is a Test?

A *test* or *inventory* in the behavioral sciences is a systematic procedure for comparing people's performance, feelings, attitudes, or values. Those tests that are most useful in research have been *standardized.* This means that the test has been published and is available, and that its questions and methods of administration are so fixed that the scores of people tested at different times and places can be compared. A standardized test usually has *norms* or *normative data,* which are statistical summaries of the performance of specified groups who have been given the test. The method for administering the test (how the test is to be given) and norms are included in the manual that accompanies the test. The manual for the State-Trait Anxiety Inventory, described at the beginning of the chapter, contains norms obtained from several hundred college students, providing a base of comparison for new data.

Test Reliability and Validity

Test reliability refers to the stability of measurement over time. When a person's typing skills are measured on two occasions, the scores should be similar unless something significant has happened in the interval between the tests. Reliability is often measured by a *reliability coefficient*—a coefficient of correlation between two administrations (see Chapter 18 for a discussion of correlation). Reliability information is generally included in the test manual along with validation information.

The *validity* of a test is the extent to which we know what it measures or predicts. Validity is judged by three types of evidence—construct, content, and criterion. *Construct validity* refers to the association of the test measurement with specific theoretical constructs. A valid personality test should be based upon or related to a major theory of personality. *Content validity* refers to the degree to which the test items reflect the domain that the test claims to cover. A test intended to measure knowledge of history should not contain items dealing with mathematics or science, unless directly related to history. Even though there might be a correlation (relationship) between scores on tests of history, mathematics, and science, due to an extraneous general intelligence factor, items in mathematics and science are generally considered outside the domain of a history examination. In recent years American courts have looked at content validity in deciding whether or not a test is discriminatory. If a particular test is used for selecting police officers, then there must be proof that the individual questions are relevant to police work. The latter decision could be based on the judgment of experienced police officers.

The third aspect, *criterion validity,* shows that the test score is related to other measures of the characteristic. A common method for determining criterion validity of a new test is to compare scores on the new test with those of earlier tests intended to measure the same characteristics. There are two subtypes of criterion validity: *concurrent validity,* the relationship between the test score and present behavior, and *predictive validity,* the ability of the test score to predict future behavior. Predictive validity has considerable practical significance in personnel selection and the identification of people at risk for certain problems who might benefit from additional assistance or counseling. To summarize, the elements of validity are:

Construct The relationship of the test to theory
Content The relevance of the items to the behavior measured
Criterion The relationship of the test to other measures
  Concurrent Correlation with existing tests or measures
  Predictive Ability to predict future behavior

Using Tests and Inventories

Technically speaking, standardized tests are less a method of research than tools, like apparatus. They are used most frequently in research in education and psychology, less often in anthropology, sociology, and geography. The researcher's use of tests is likely to be different from their use in career planning, counseling, or personnel selection. Some of the most commonly used standardized tests are IQ tests, personality tests, occupational inventories, and school achievement tests (Box 15-1). There are also specialized instruments such as self-esteem measures, sex-role inventories, environmental rating scales, and other topical tests used in specific areas of research.

BOX 15-1. Common Types of Standardized Tests in Educational and Psychological Research

Test type	*Purpose*	*Examples**
School achievement	To determine a person's academic level	Metropolitan Achievement Test, Stanford Achievement Test
Intelligence (group administration)	To measure how a person's intelligence compares with that of others in the same age group	Henmon–Nelson Tests of Mental Ability, SRA Primary Mental Abilities
Intelligence (individual administration)	To obtain a detailed picture of an individual's intellectual abilities, primarily for clinical use	Wechsler Adult Intelligence Scale, Stanford-Binet Intelligence Scale

Personality	To measure various dimensions of normal personality, such as introversion, independence, practicality, etc.	California Psychological Inventory, The Personality Inventory
Mood	To assess emotional state	POMS (Profile of Mood States), the Adjective Check List (Gough), STAI (State-Trait Anxiety Inventory)
Clinical	To identify patterns of behavioral dysfunction	MMPI, IPAT Depression Scale
Occupational interests	To determine interest in specific vocations as well as general occupational categories	Strong-Campbell Interest Inventory (SVIB-SCII), Ohio Vocational Interest Survey
Specialized abilities	To measure abilities in such areas as music, art, languages, etc.	Seashore Measures of Musical Talents, Modern Language Aptitude Test, General Clerical Test

*Only a few tests in each category are listed here. Many others can be found in the catalogs of test distributors.

Some tests are designed for group administration, while others are constructed for individual administration. Most group-administered tests can be given by a relatively untrained person and scored by computer. The interpretation of the results (what the scores on the various parts of the test mean) must be done by a knowledgeable person. Other tests are designed to be individually administered, scored, and interpreted by a trained examiner.

Personally constructed instruments, such as a questionnaire used for a single project or an examination given by a college instructor, do not qualify as standardized tests, but with sufficient development, revision, and the compilation of norms, these instruments could become standardized tests. The advantage of using a standardized test is that someone else has done the work. You need not develop your own measure of hypnotizability or employee morale; there are tests already available. Booklets and answer sheets are printed and norms allow comparison of new scores with those of others who have taken the test. Some tests are accompanied by test keys that allow you to score the answers yourself; others must be mailed to a testing service, where they are scored and analyzed for a fee.

Many standardized tests are culture-specific. An intelligence test developed in one nation may not be useful or valid in another. There may be serious problems with some of the questions or tasks required. This does not mean that international norms for standardized tests are impossible to achieve. It does mean that further examination and administration is necessary before a test can be assumed valid in another country. For example, when researchers in Italy translate psychological tests developed in Germany, they must standardize the tests on an Italian sample before they assume validity for use in Italy.

Some tests have multiple forms for use with different groups and nationalities. As an example, the Sixteen Personality Factor Questionnaire (16PF) has norms for use with different age groups, including a preschool version, an early school version, a children's form, a secondary school form, and a version for those 16 years and older. There are German and Spanish language editions, and test forms in Braille and on videotape using sign language.

Unfortunately, standardized tests are often unsuitable for your specific needs. Very few of the research problems discussed in previous chapters are suitable for standardized testing. No commercial questionnaire will tell you what college students think about the cafeteria food; the same is true of student attitudes toward peer counseling, financial aid, or alcohol use. Few standardized instruments with appropriate norms are available for attitude measurements. Even those standardized tests on which the most work has been done, such as measures of intelligence and school achievement, have many limitations and may not be usable for the particular groups you want to test. However, if you decide that the standardized instrument comes close to your needs, do not tinker with it. If you change the procedure or reword the items, you will lose the advantage of having the normative data (usually presented in the test manual) with which to compare your findings. Also, your results will have more meaning if they can be compared with the findings of other researchers using the same instrument in the same manner.

Locating Suitable Tests

Your first task is to find out whether appropriate tests are already available. This can be done in several ways. First, you can look at previous research on the topic. Have studies similar to yours employed standardized tests? If not, the answer may be that none is available.

A very comprehensive index is *Tests in Print (TIP),* providing a bibliography of all known commercially available English tests in print. *TIP III* has 2672 test entries. It serves as an index for the *Mental Measurements Yearbook (MMY).* The latter volume provides test reviews and evaluative information, as well as a reference list to relevant articles. The most recent is the tenth edition. If that is not available in your library, use an earlier edition.

Another source is Sweetland and Keyser's *Tests: A comprehensive reference for assessments in psychology, education, and business.* There is a subject index and a brief description of each test. That volume is accompanied by several volumes titled *Test Critiques* that provide excellent descriptions of many tests. Both *TIP* and *Tests* provide price information and the names and addresses of test suppliers.

Ask knowledgeable individuals whether they know of tests or inventories appropriate for your purposes. Faculty members in departments of education and psychology, personnel managers, school guidance counselors, and others who use standardized tests in their work, may have catalogs from test suppliers. It is not easy for a beginner to sort through numerous catalogs and find the right test. Discussing your needs with a qualified test specialist is helpful.

Obtaining the Tests

Some tests are available only to members of professional societies, such as the American Psychological Association. Access to these tests is restricted to prevent their unauthorized use. Information on availability is contained in most test catalogs. If the test you are considering has restricted distribution, you must either be personally qualified or involve someone who is. The personnel department of your company or the counseling bureau of your school probably has a qualified test professional. This person may have a copy of the test on file or can obtain a sample copy for you to examine. Most test suppliers distribute (for a small charge) sample kits containing a single copy of the test booklet, a test manual containing norms, and scoring sheets. The sample kit should be closely scrutinized before any decision is made about ordering multiple copies. You may find that the questions on the test are too easy, too difficult, or inappropriate for the sample in your study.

Constructing a Test

If you cannot locate the test you need in catalogs from test suppliers or published work on the topic, a remaining option is to develop a test of your own. This will be a time-consuming procedure so the decision should not be made lightly. Making up a standardized test, with score sheet, norms, etc. requires far more time and effort than composing a questionnaire or list of interview questions.

The sequence of steps involved in compiling a standardized test is shown in Box 15-2. The first task is to define the performance or behavior to be measured. This is a more difficult task than it may sound, because it involves important policy decisions. Let us use bicycle riding as an example of a common ability for which few or no standardized tests are available. If you wanted to compare the bicycle riding skills of individuals or groups, you would probably have to make up your own test. Defining the skill will raise

BOX 15-2. Steps in Constructing a Standardized Test

1. Define the performance to be measured
   Check previous research
   Include your own experience
   Consult with recognized authorities
   Translate statements into behavioral measures
2. Compile potentially useful items
   Eliminate duplications and inconsistencies
   Decide on length and format
3. Pilot test
   Revisions (several)
4. Compose final version
   Item analysis
5. Check reliability by administering test to same individuals on two occasions
6. Check validity by comparing scores with known abilities or performance
7. Compile norms
   Administer tests to different groups
   Compile scores according to relevant categories (e.g., age, education, experience, etc.)
8. Publish or disseminate test

some important questions. Should you include items pertaining to mountain bikes and trail riding? What about skills for long distance touring or racing? Or riding a unicycle? These are certainly "skills," but they may not be particularly relevant if your concern is with safety in local bicycling. One approach to defining the topics to be included in a test is to consult previous research. What aspects of performance have been deemed important or relevant by other investigators? Another possibility is to consult with recognized authorities. What sort of skills do leaders of cycling clubs and safety officials feel are important for bicyclists? General statements must be made specific and defined in terms of observable operations.

The behavioral items collected from all sources must be examined to remove obvious duplications and inconsistencies. See that important issues are covered and that the instrument is not too long. The revised list of items can be tried out on an exploratory basis with a small number of respondents. The goal at this point is to get the bugs out of the procedure. This will require several successive revisions of the preliminary version of the test. This will produce an almost-final version of the test to be further refined through an *item analysis.* Detailed methods for undertaking an item analysis can be found in Murphey and Davidshofer (1988). An item analysis shows the degree to which the various items "hang together." Items that produce responses unrelated to those of the other questions are probably measuring a different type of ability and can be removed from the instrument.

Establishing the reliability of the revised instrument is done by administering the test to the same individuals on two separate occasions. If the test is reliable, those individuals who score high on the test the first time should also score high on the second administration. Many standardized tests will also have alternate versions. Constructing alternate versions requires additional effort but it will increase the practical and research usefulness of the instrument.

One approach to validity is to administer the test to people known to be high or low in the skill being measured. Presumably a group of experienced cyclists should perform better on a test of bicycle skills than a group of inexperienced cyclists. Another approach to validation could test individuals who have been involved in bicycle accidents. We could predict that when matched for age, cycling experience, amount of riding per week, and other relevant background factors, individuals who have been involved in a bicycle accident should perform worse on a test of bicycle skills than individuals who have not been involved in an accident. These are only a few of the ways in which a standardized test can be validated.

The next step is to compile norms from various groups. A test of bicycle skills should probably include norms for different age groups. The norms would be compiled by administering the test to samples of respondents at different age levels and tabulating the results. This will allow scores of people to be compared with those of their own age groups.

If all of the preceding steps have been done, and the test appears promising in predicting bicycle skills, the next step will be to publish the results and make them available to researchers and cycling organizations. Encouraging researchers to try out the test will assist in developing norms for other geographic areas and types of samples.

Projective Tests

Tests that have not been standardized are sometimes useful in research. The absence of norms or score sheets may mean that the test is new and the investigator wants to try it out. Some instruments, which rely on subjective interpretation by a trained examiner, do not lend themselves to standardization. There is a class of research techniques in clinical psychology called *projective tests* that lacks usable norms. They are called "projective" because the questions or stimuli are deliberately vague or incomplete. It is assumed that the respondents fill in the blanks by *projecting* their personal concerns and experiences. Often the style of responding (e.g., hesitation or embarrassment) is more important than the content of the response. The two most widely used projective tests are the Rorschach and the TAT (Thematic Apperception Test). The Rorschach consists of 10 inkblots developed by the Swiss psychiatrist Herman Rorschach. This was the successor to the Stern Cloud Pictures Test, in which the respondent was shown pictures of clouds and asked what they looked like. Multiple-choice versions of the inkblot test suitable for group administration are also available. The TAT consists of a set of blurred pictures of people in various situations. It was developed at the Harvard Psychological Clinic by Christina Morgan and Henry A. Murray.

Limitations

Standardized tests are not available for many topics that an investigator wants to study. When tests are available, the norms may not be appropriate for groups other than those on which the test was standardized. Many standardized tests have been criticized as being culturally biased. Their norms are likely to be inappropriate for people who speak another language, have been out of school a long time, or who are institutionalized. Standardized tests are primarily useful for people similar to those on whom the norms were standardized. The apparent precision of test scores is often misleading. The important issue is how well the test measures what it is supposed to measure.

Personality tests tend to stigmatize people who diverge from conventional social norms. Some of the questions on personality tests are intrusive. You must be sensitive to the way the question appears to respondents and outside agencies. It is difficult to defend some of the questions about personal habits and sexual preferences included on major personality tests.

The research value of all projective tests is limited by their low reliability. Two psychologists examining the same record are likely to develop different interpretations.

Summary

Standardized tests, which are available from test supply firms, have norms that can be used in interpreting results. *Reliability* refers to the stability of measurements over time. Three aspects of test validity are *construct,* the relationship of test to theory, *content,* the relevance of the items to the behavior measured, and *criterion,* the relationship of the test to other measures (includes *concurrent* and *predictive* validity). Standardized tests tend to be culture-specific. The most common types measure school achievement, intelligence, personality, clinical symptoms, occupational interests and skill, and specialized abilities in areas such as music, art, and languages.

Information about tests and their availability can be found in *Tests in Print,* and *Tests,* volumes that are available in most reference libraries. Constructing a new standardized test is a difficult and time-consuming task because of the need to establish reliability and validity, and to collect norms from different segments of the population.

Nonstandardized tests may be so new that norms have not been developed. Or they may be deliberately vague, as with the projective tests used in clinical psychology, so the subject can give an individual interpretation to the questions.

Limitations: Many standardized tests are not appropriate for groups other than those on whom the norms were established.

References

Amercian Educational Research Association, American Psychological Association, National Council on Measurement in Education (1985). *Standards for educational and psychological testing.* Washington, DC; Authors.

Conoley, J. C. & Kramer, J. J. (eds.) *The Tenth Mental Measurements Yearbook.* Lincoln, NE; University of Nebraska Press.

Keyser, D. J. & Sweetland, R. C. (eds.) (1984–1988). *Test Critiques* (Vols. 1–7). Kansas City: Test Corporation of America.

Mitchell, J. V., Jr. (1983). *Tests in Print III: An index to tests, test reviews, and the literature on specific tests.* Lincoln, NE: University of Nebraska Press.

Murphy, K. R. & Davidshofer, C. O. (1988). *Psychological testing: Principles and applications.* Englewood Cliffs, NJ: Prentice Hall.

Speilberger, C. D., Gorsuch, R. L., & Lushene, R. E. (1970). *Manual for the State-Trait Anxiety Inventory.* Palo Alto, CA: Consulting Psychologists Press.

Sweetland, R. C. & Keyser, D. J. (eds.) (1986). *Tests: A comprehensive reference for assessments in psychology, education, and business* (2nd ed.). Kansas City: Test Corporation of America.

Further Reading

Cronbach, L. J. (1990). *Essentials of psychological testing (5th ed.).* New York: Harper & Row.

Crocker, L. M. (1986). *Introduction to classical and modern test theory.* New York: Holt, Rinehart and Winston.

Morris, L. L., Fitz-Gibbon, C. T., & Lindheim, E. (1987). *How to measure performance and use tests.* Beverly Hills, CA: Sage.

Murphy, K. R. & Davishofer, C. O. (1988). *Psychological testing: Principles and applications.* Englewood Cliffs, NJ: Prentice Hall.

Walsh, W. B. (1989). *Tests and measurements* (4th ed.). Englewood Cliffs, NJ: Prentice-Hall.

VI
Sampling, Statistics, and Reports

16 Sampling

A physical education teacher wants to study the possibility of sex differences in running speed among elementary school children. This will involve a comparison of the running time of boys and girls in different age categories. For practical reasons, the teacher is unable to test all the children in the school. This is a common situation in behavioral research. Rarely can an investigator test all the people in a given category. Some selection of subjects is necessary, but this must be done in an unbiased way. The entire group of people in a category is called a *population.* The smaller group selected for testing is called a *sample.* The sample is then used to make generalizations about the population from which it was drawn.

The degree to which the sample differs from the population is termed *error.* There are two general sources of error: sampling error and sample bias. *Sampling error* refers to chance variations among samples selected from the same population. Imagine a person who draws successive samples of 50 marbles each from a large bowl containing 1000 marbles, half black and half white. After each selection is noted, the marble is returned to the bowl. Some samples would contain a majority of black marbles and others a majority of white marbles. Ideally, if there were enough trials, the characteristics of all the samples would *average* 25 white marbles and 25 black marbles. However, any single sample of 50 marbles is likely to deviate from this average. The difference between the samples and the actual population characteristics (50 percent black and 50 percent white) is the *sampling error.* This can never be totally eliminated, but it can be estimated statistically. A common method of reducing the influence of chance variation is to draw a large sample. As the sample size approaches the population size, it becomes a more accurate representation.

Sample bias refers to error introduced by a sampling procedure that favors certain characteristics over others. In selecting the marbles from the jar, it is possible that the black marbles, by absorbing heat, are slightly warmer than the white marbles. The blindfolded subject may unconsciously select the warmer black marbles. This source of error could be reduced by having the blindfolded subject wear gloves. Increasing the sample size is not effective in reducing sample bias. The source of the bias must be identified and then reduced or eliminated.

Types of Samples

Although probability sampling, described below, is necessary for generalization to the population (a matter of external validity), there are occasions when we do not need to go through the trouble and expense of obtaining a probability sample. For example, we might have a specific group of people in mind, or the study is a class project where learning how to interview is more important than generalizing the findings. In such cases, *nonprobability* sampling is acceptable. The specific type will depend on the needs of the project.

Probability Sampling

Probability samples are those in which we know the probability for the inclusion of any given individual. Two general types of probability samples are *random samples* and *stratified samples.* Visiting a college campus and interviewing every fourth person on the sidewalk is *not* likely to yield a random sample of the student body, because each student will not have an equal chance of being selected. The chances of being selected will be affected by the location, time of day, weather, and so on.

Random Sample

In random sampling, every person in the entire population being studied has an equal likelihood of being selected. For random sampling, the researcher must have access to each member of the population, that is, know how many people there are and how to reach them.

Drawing names out of a hat is one way of selecting students randomly from a class. A more formal procedure is to use a table of random numbers, such as Table A-5 in Appendix A. These are numbers that follow no particular pattern. The researcher can start anywhere in the table and proceed in any direction.

Example

The gym teacher interested in running speed wants to select a random sample of 20 first-grade boys and 20 first-grade girls from a school population of 71 girls and 68 boys in the first-grade classes. The teacher obtains a combined list of all first-grade classes. Then, using a table of random numbers, the teacher begins anywhere on this table, assigning successive two-digit numbers to every boy and girl on the list. For this particular study, the teacher started at the line marked in the second column and proceeded down the list.

Student name	*Two-digit numbers from Table A-5*
Jim A.	72
Sue F.	09
Bill S.	43
Martin F.	48
Alice P.	90
Nancy M.	61
.	.
.	.
.	.

The researcher can then select the 20 boys and 20 girls with the lowest numbers for the sample. If some of these students cannot participate, additions to the sample can be made from among those with the next lowest numbers.

A simpler method for selecting a sample, but with a higher risk of bias, involves the use of every third or fourth name on the list. The use of the first 20 names should be avoided unless the list has been compiled in a fully random manner. Students who registered late may be at the end of the list; some ethnic groups will be more heavily represented in some parts of the alphabet than in others. These same considerations apply in selecting locations in which to interview people. Houses for a neighborhood survey should be chosen according to a random procedure. Allowing the interviewer to select houses may introduce sampling bias (e.g., choosing only the more attractive houses and ignoring those that are rundown or have "No Trespassing" signs). Corner locations may differ in some ways from those in the middle of the block (more traffic? higher real estate values?). The way to avoid such problems is to develop a clear procedure for randomly selecting houses rather than leaving the decision to the interviewer.

Stratified Sample

A stratified sample is a variation of the random sample. Instead of selecting each person randomly, criteria are set up to ensure representation of particular groups within the sample proportionate to their numbers in the population. For example, if the population in question is made up of 55 percent males and 45 percent females, then a stratified sample is set to have 55 percent males and 45 percent females. Similar criteria may be set for age, residence, religion, and other variables of interest. Within categories, individuals are chosen randomly. Public opinion polls rely on stratified samples. They select a relatively small number of families to represent various levels of age, income, region, ethnic status, and other variables. The use of a representative sample requires an accurate knowledge of the characteristics of the population.

Nonprobability Sampling

Samples in which the likelihood of selection is not actually known are called nonprobability samples. There are three general types: *quota samples, purposive samples,* and *accidental samples.*

Quota Sample

Sometimes the researcher does not want or need a perfectly stratified sample. Let us imagine that in the school in which the running study is being conducted, there is only a small number of Hispanic students. If the researcher draws a random or stratified sample, it will probably provide too few Hispanic students in each sex and grade level. An alternative is for the researcher to use a *quota sample* in which the selection categories are specified according to the needs of the investigator. Within those categories, individuals are chosen randomly. A quota sample of school children might include 50 percent boys and 50 percent girls, one-sixth of them at each grade level, and one-third from each major ethnic category (black, white, Hispanic). These percentages would *not* necessarily reflect the actual composition of the school. They would, however, provide enough individuals in each part of the research design to permit adequate comparisons to be made. While the number of individuals in each category of a quota sample is chosen according to the researcher's needs, the particular individuals are chosen randomly.

Purposive Sample

A *purposive sample* is one in which individuals thought to be most important or relevant to the issue studied are targeted for the research. For example, a study of city policy may require interviews with particular individuals in leadership positions. For an experiment studying the effect of two types of video terminal monitors on employee productivity, one might select the employees who will be using the system most often to be the experimental and control subjects.

A special type of purposive sample is the *snowball sample* where the researcher asks respondents for other persons to contact. This technique is useful when studying particular groups where membership may not be obvious or where access to members may be difficult, for example, political activists, members of a religious cult, or gang members.

Accidental Sample

Accidental sampling is taking what you can get. Interviewing people on a particular day in a local shopping mall, or observing students at lunchtime on campus are examples of accidental samples. A variant is the volunteer sample—comprised of people who are willing to participate in a research project. The characteristics and behavior of volunteers may be quite different from those of nonvolunteers.

Generalizability

Generalization from the results obtained with a probability sample is limited to the population from which the samples were drawn. Elementary schools differ so much that it is risky to generalize from one school to another. The investigator can subsequently enlarge the study to include students from schools in other parts of the country or in other nations. If such studies repeatedly turn up the same sex difference, this result can reasonably be considered genuine and not simply characteristic of a single school.

If the investigator has used a nonprobability sample such as a quota plan that does not reflect the population's characteristics (randomly choosing 100 black, white, and Hispanic children when this is not the ethnic ratio among the student body), generalization can be made only within each ethnic group. It cannot be extended to the total student body unless some adjustment is made based on the proportional membership of each ethnic group in the total population.

While nonprobability samples are limited in generalizability, steps can be taken to increase their validity. Purposive or quota samples should be selected randomly within the targeted groups. Observations made at different times of day will reduce some bias. Interviews in a shopping mall can be made of every fifth person entering. Using systematic procedures that are devised ahead of time may be very effective in reducing sample bias even when a pure random sample is not obtained. The point is to try to approximate randomness in the technical sense (to provide as equal an opportunity for selection as is possible), and to devise ways of reducing or eliminating sources of error.

How Large a Sample?

Other things being equal, large samples provide more reliable and representative data than small samples. Samples of 50 marbles randomly selected from the 1,000-marble bowl will show less variability in the proportion of black and white marbles than will smaller samples of 10 marbles each. While it would be unusual to select 40 black and 10 white marbles, it would be far more likely to draw a sample of 10 marbles that contained 8 black and 2 white marbles.

In deciding on sample size, there is an interesting difference between statistical logic and practicality. The reduction in sampling error that results from a large sample may not be worth the extra time and effort required to obtain the additional data. Some public opinion polling firms find that they can make accurate predictions of voting trends based on interviews with a representative sample of 2500 individuals. To increase the sample to 5000 or 10,000 would reduce sampling error but would not be worth the extra time, effort, and expense. In public opinion surveys, time is very important. The extra two days that might be required to double the sample would make the results less useful to the client.

The following factors should be considered in making a decision on sample size:

1. *Size of population.* For a survey of library *readers,* a sample of 100 individuals would probably be sufficient. For a survey of library *employees,* this would be far too many. There may not even be 100 employees. The researcher might eventually decide upon a 50 percent sample among library desk clerks and an 80 percent sample of administrators in order to obtain enough individuals in each category.

2. *Available resources and time constraints.* Pilot testing will reveal the cost in terms of time and effort for each testing session. The *maximum* size of the sample can be determined from such constraints. The *optimal* size of the sample will depend on other factors.

3. *Strength of the effect.* When the independent variable has a strong and clear effect on the dependent variable, a smaller sample can be used. For example, only a few subjects would be needed to demonstrate the effect of five ounces of alcohol on reaction time. A much larger number of subjects would be necessary to determine the effect of one ounce of alcohol, since this produces inconsistent and often contradictory results.

4. *Number of subsets to be compared.* If the researcher wishes to compare groups within the sample by dividing it according to social class, age, ethnicity, or gender, for example, the sample must be large enough to include a sufficient number of individuals in each of the subcategories.

5. *Refusal and spoilage rates.* A sample must be increased to allow for unusable data. It may be possible to predict in advance the number of people who will be unable to follow the instructions, drop out of the experiment, or terminate the interview prematurely. A questionnaire given to and collected from a group of clerical employees is likely to have a high return rate, perhaps in excess of 80 percent. However, the same

questionnaire mailed to an unselected group of individuals with a request to return by mail may have a return rate of 10–15 percent.

Sample size should be specified in advance. This avoids the accusation that data collection was halted as soon as the results supported the hypothesis. The completion of one study does not rule out the possibility of additional ones. It is useful to begin a study with a small sample. The investigator interested in sex differences in running speed might compare 20 boys and 20 girls in one class. If the trends are very clear, the researcher could stop testing in that grade and look for trends in other grades. However, if the original results indicated a weak trend, the researcher could repeat the study with additional groups of boys and girls at the same grade level. Two independent studies represent a more powerful test of a hypothesis than a single study involving twice as many subjects.

A description of the sampling procedure should always be included in the survey report. Here is an account of the methods used in a telephone survey of consumer attitudes toward food preservatives:

> A questionnaire developed for a survey of 370 consumers from the Statistical Metropolitan Survey Area of Seattle-Everett, Washington, in 1974 was adapted for use in this survey. Structured interviews were conducted in the same district by telephone. Phone numbers were selected at random from the telephone directory; 220 households were contacted and 170 usable replies were obtained. Only persons responsible for the major food purchases for the household unit being studied were interviewed. Each interview took 5–7 minutes, and all were conducted by the same interviewer (Martinsen & McCullough, 1977).

Limitations

A sample that is perfectly representative is difficult to obtain. Census records are probably out of date, not everyone is listed in the telephone directory, and door-to-door surveys are expensive. Many investigators are willing to settle for less than a perfect sampling procedure in order to obtain samples that are accessible and cooperative.

Summary

The entire group of people in a category is called a population. The smaller group selected for testing is called a sample. The degree to which the sample differs from the population is termed error. Two general sources of error are sampling error and sample bias. Sampling error refers to chance variations among samples selected from the same population. It is reduced by increasing the size of the sample. Sample bias refers to error introduced by a sampling procedure that favors certain characteristics over others. Ways of reducing sample bias are using probability samples or eliminating the bias by good research design such as using systematic procedures where the sample cannot be truly random.

Probability samples are either random or stratified. In a *random sample,* each member of the population has an equal chance of being selected. A *stratified sample* is selected so that its characteristics are proportionate to those present in the total population.

There are three types of nonprobability samples: quota, purposive, and accidental. In a quota sample, the selection categories are specified according to the needs of the investigator. A purposive sample targets key individuals for study. An accidental sample uses what's available.

The generalizability of a probability sample is limited to the population from which it is drawn. Nonprobability samples have very limited generalizability (external validity).

Other things being equal, large samples provide more reliable and representative data than small samples. The decision on sample size is influenced by the size of the population, available resources and time constraints, strength of the effect, number of subsets to be compared, and expected refusal and spoilage rates.

Limitations: Sampling is often done under less than ideal conditions. An investigator may be willing to trade off some representativeness for a sample that is more accessible and willing to participate in the study.

Reference

Martinsen, C. S., & McCullough, J. (1977). Are consumers concerned about chemical preservatives in food? *Food Technology,* 7, 56–59.

Further Reading

Backstrom, C., & Hursh-Cesar, G. (1981). *Survey research.* New York: Wiley (see Ch. 2).

Chein, I. (1976). An introduction to sampling. In C. Selltiz, L. S. Wrightsman, & S. W. Cook (Eds.), *Research methods in social relations* (3rd ed.) (pp. 511–540). New York: Holt, Rinehart and Winston.

Kraemer, H. C., & Thieman, S. (1987). *How many subjects?* Beverly Hills, CA: Sage.

Sudman, S. (1983). Applied sampling. In P. H. Rossi, J. D. Wright & A. B. Anderson (Eds.), *Handbook of survey research* (pp. 145–194). New York: Academic Press.

17 Descriptive Statistics

How many uses can you think of for a top hat?

"It could be a scoop, a target, a container, a musical instrument."

"You could throw snowballs at it, use it as a paperweight, a mouse might use it for a boat on the water. You could use it as a fan if you got hot."

"You could use it as a vase or keep sugar or fish in it."

These are some of the answers students have given to the top hat question, part of a test used to measure creativity. Students receive a score for every new category of uses. The student who said that a top hat could be used as a vase, a sugar bowl, and fishbowl would receive only one point, since these are all single uses of the top hat (i.e., as a container). Table 17-1 shows a set of scores for 16 college students asked the top hat question.

These figures illustrate the *raw data* for the test. The raw data consist of the firsthand test scores. Notice how difficult it is to make any sense out of all these replies. Here a knowledge of statistical technique is useful. Statistics provide a way to reduce a mass of numbers to manageable proportions. This process is neither mysterious nor frightening; in general, it follows the rules of common sense. The best way to learn statistical techniques is to apply them to a set of data. In much of this chapter, we will be using the creativity test scores presented in Table 17-1. The measure used is a continuous one. Continuous measures are scores that vary along a continuous dimension or distribution, such as age in years, months, and days; income in dollars and cents; and creativity test scores ranging from zero to some upper limit. The following descriptive statistics are used for *continuous measures.* They do not apply to *categorical* measures—counts of separate and distinct units.

TABLE 17-1. Number of Uses for a Top Hat Given by 16 College Students

Student	*Number of uses*
1	8
2	5
3	5
4	11
5	6
6	7
7	6
8	4
9	9
10	4
11	5
12	7
13	6
14	6
15	8
16	7

Sex, country of residence, and political party affiliation are examples of categorical measures. They do not reflect continuous dimensions. A person is scored as falling into one category or another.

Frequency Distributions

An initial step in summarizing test results is to arrange them in a frequency distribution. This is a table with two columns. The scores from highest to lowest are written in one column; the *frequencies,* or number of times each score occurred, are in the second column.

Commonly used symbols or abbreviations in a frequency distribution are:

X = raw score
f = frequency
N = number of cases

Number of uses for a top hat X	Number of people giving this number of uses f
11	1
10	0
9	1
8	2
7	3
6	4
5	3
4	2

When the scores span a wide range, they can be grouped in *intervals* of equal size. Intervals may be of any size, although multiples of five are the most common. Interval grouping works best with 10–20 intervals. Fewer than 10 will obscure the differences in spread, while too many intervals are likely to dilute any trends. Because there are only 16 creativity test scores and since they vary over such a limited range (4 to 11), there is no point in interval grouping. Table 17-2 uses a different set of test scores to illustrate the use of interval grouping in a frequency distribution.

TABLE 17-2. Frequency Distribution of Arithmetic Achievement Scores for 142 Children, Ages 10 to 10½ Years

X	*Tally*	f
160–169	|	1
150–159	|	1
140–149	|||	3
130–139	𝍸 𝍸 ||||	14
120–129	𝍸 𝍸 𝍸 ||||	19
110–119	𝍸 𝍸 𝍸 𝍸 𝍸 ||	27
100–109	𝍸 𝍸 𝍸 𝍸 𝍸 𝍸 ||	32
90–99	𝍸 𝍸 𝍸 𝍸 |||	23
80–89	𝍸 𝍸 𝍸	15
70–79	||||	4
60–69	||	2
50–59	|	1

Note: The *tally* should not be included in your presentation of results. It is shown here merely to illustrate how the frequencies are obtained.

Cumulative Frequencies

Cumulative frequencies are shown in Table 17-3. These are formed by beginning at the bottom of a frequency distribution and summing all the frequencies at and below each level. An advantage of a cumulative distribution is that it shows the number of cases falling below each interval score. From the third column in Table 17-3, we find that 45 children scored below 100 (45 were at or below the interval of 90–99). Note that the cumulative frequency listed at each interval includes those cases within the interval as well as those below. The cumulative frequency can be transformed into a *cumulative percentage* by dividing it by the total number of cases and multiplying by 100. The 45 children scoring below 100 are 31.7 percent of the total sample (see Table 17-3, fourth column).

Percentile and Percentile Rank

Assume that your cousin applied to college and took a standardized entrance examination. The college of her choice seeks to admit only the top 10 percent. She looked at the published norms and found that 90 percent of those having taken the test scored 260 or lower. In this example, 260 represents the ninetieth percentile. She has to score above 260 on the test. A *percentile* is the score at or below which a given percentage of cases lie. In a frequency distribution the cumulative percent shows the percentile value.

TABLE 17-3. Arithmetic Achievement Scores for 142 Children, Ages 10 to 10½ Years, Expressed in Cumulative Frequencies and Percentages

Interval	*f*	*Cumulative f*	*Cumulative %*
160–169	1	142	100.0
150–159	1	141	99.2
140–149	3	140	98.6
130–139	14	137	96.4
120–129	19	123	86.7
110–119	27	104	73.2
100–109	32	77	54.2
90–99	23	45	31.7
80–89	15	1 + 2 + 4 + 15 = 22	15.5
70–79	4	1 + 2 + 4 = 7	$\frac{7}{142} = 4.9$
60–69	2	1 + 2 = 3	$\frac{3}{142} = 2.1$
50–59	1	1	$\frac{1}{142} = 0.7$

Checking back with your cousin, you find that she received a score of 251 on the entrance examination. Although she wasn't admitted to the college of her choice, she wonders about her relative performance on the test (i.e., how she stood in relation to the other students). One method of evaluating her score is to find its *percentile rank.* This is the percentage of cases in the comparison group at or below the score in question. In the case of your cousin, it is the percentage (without using the percent sign) of people receiving her score or less. If 82 percent of the test takers score 251 or lower, the percentile rank of 251 is 82. To calculate percentile rank, use the following formula:

$$\frac{\text{Number of cases at or below a given score}}{\text{Total number of cases}} \times 100 = \text{Percentile rank}$$

Figures and Graphs

Up to this point, the data have been presented in statistical tables. It is also possible to present the same data in graphs and figures. One form of presentation is the *bar graph* or *histogram* (see Figure 17-1). Here the intervals are shown along the horizontal axis (base) and the frequencies along the vertical axis.

The *frequency polygon* is another form of graphic representation. As in the bar graph, the individual scores or the intervals are listed along the hor-

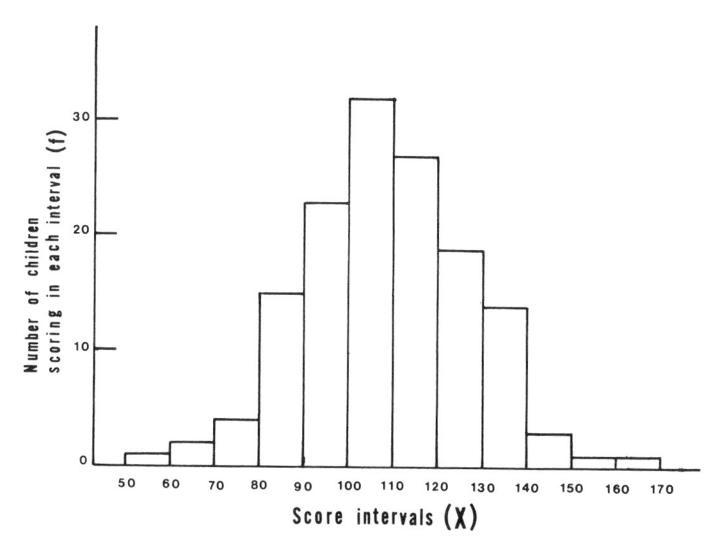

FIG. 17-1. Bar graph showing arithmetic achievement scores for 142 children, ages 10 to 10½ years.

izontal axis. However, instead of bars, dots are placed on the graph at the *midpoint* of each interval to indicate the frequency of scores. Then a line is drawn connecting the points (Figure 17-2).

When comparing different-sized groups, bar graphs and frequency polygons can be based on percentages instead of frequencies. Cumulative frequencies and cumulative percentages can also be presented in graphic form. Figure 17-3 shows cumulative percentages for the same data as both a bar graph and a curve.

While the rules for constructing graphs and figures are not rigid, there are some guidelines:

1. Scores are presented along the base (horizontal axis) and frequencies or percentages on the vertical axis.
2. Numbers run from left to right, beginning with the low numbers. Low numbers are placed at the bottom of the vertical axis.
3. The entire graph and everything on it should be clearly labeled. Any symbols used should be defined.
4. The overall impression conveyed by a graph is affected by the ratio of the vertical to the horizontal axis. Some writers suggest keeping the vertical axis about three-quarters as long as the horizontal axis.
5. The title of a table or figure is called the *legend.* This is generally placed above tables and below figures and graphs.

This may be the time to stop momentarily. The next part of the chapter will deal with measures of central tendency. However, before going on, you may wish to review some of the earlier parts of this chapter.

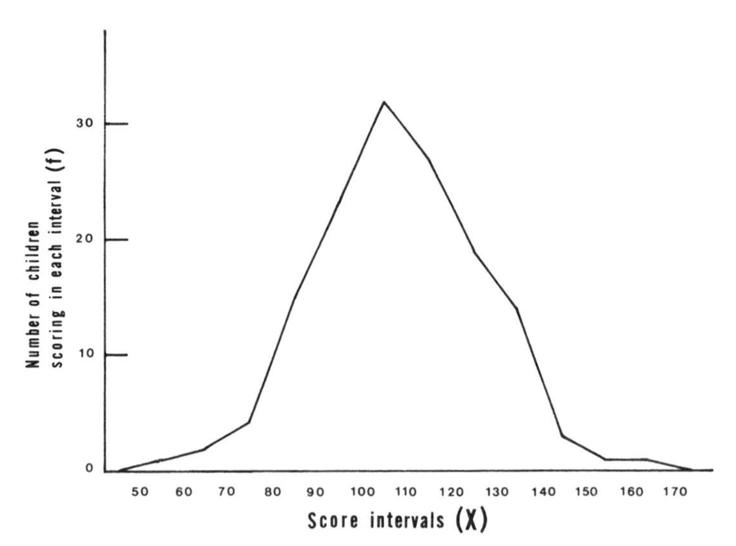

FIG. 17-2. Frequency polygon showing arithmetic achievement scores for 142 children, ages 10 to 10½ years.

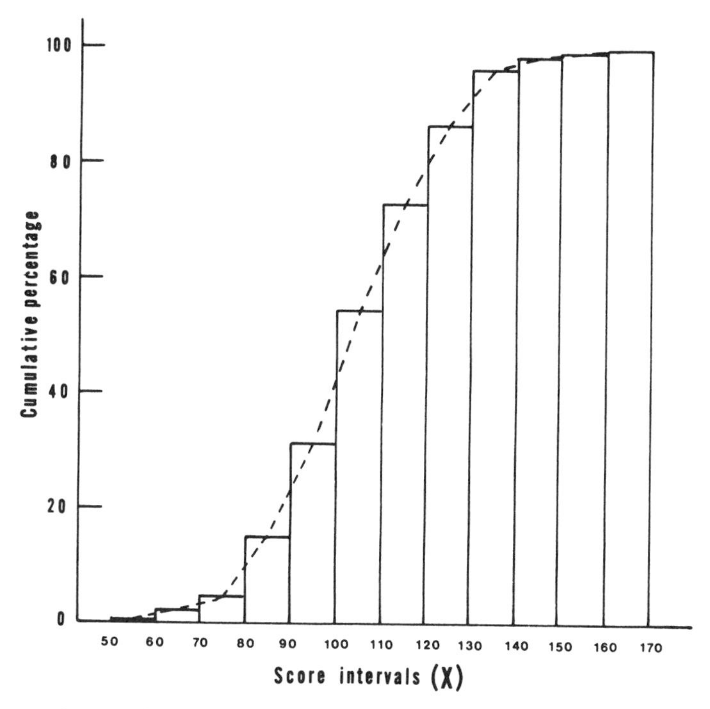

FIG. 17-3. Cumulative percentages of arithmetic achievement scores for 142 children, ages 10 to 10½ years.

Measures of Central Tendency

A further step in the analysis of continuous scores is the calculation of an average. There are three types of averages: mean, median, and mode. If you have taken a test, knowing the mean, median, and mode gives you a better idea of your performance relative to that of others who took the test. Measures of central tendency are also useful for comparing the scores from two groups of people on the same test.

Mean

The mean is the arithmetic average. There are two ways of calculating it. (1) Add up the individual scores and divide by the total number of cases.

Formula: $\overline{X} = \dfrac{\Sigma X}{N}$

where $\overline{X}$ = mean
Σ = sum of
X = raw score
N = number of cases

Example
The mean of the creativity scores is calculated for the test scores as follows:

Sum of all scores (ΣX) = 104
Number of test scores (N) = 16

$$\text{Mean } (\overline{X}) = \frac{104}{16} = 6.5$$

When data are in a frequency distribution:

Multiply each score X by its frequency,
Sum those products,
Divide by the total number of cases.

$$\text{Formula: } \overline{X} = \frac{\Sigma(Xf)}{N}$$

Table 17-4 illustrates the calculation of the mean from the frequency distribution of creativity scores.

The mean is the most frequently used measure of central tendency. It is usually the most reliable of the three measures since it varies less among samples drawn from the same population.

Median

The median is the midpoint of the distribution when all the scores are arranged from highest to lowest. Half of the scores fall above the median

TABLE 17-4. Calculation of the Mean from the Frequency Distribution of Creativity Scores

Number of uses for a top hat X	*Number of people giving this number of uses* f	*(X multiplied by f)* Xf
11	1	11
10	0	0
9	1	9
8	2	16
7	3	21
6	4	24
5	3	15
4	2	8
	$N = 16$	$\Sigma(Xf) = 104$

$$\text{Mean } (\overline{X}) = \frac{\Sigma(Xf)}{N} = \frac{104}{16} = 6.5$$

and half below. The advantages of the median over the mean are that it is quickly calculated from a frequency distribution, and it is a better indicator of central tendency when there are a few extremely high or extremely low scores. The mean is much more influenced by extremes than is the median.

To calculate the median from a frequency distribution, arrange the scores from highest to lowest. The median is the middle score, or the point between the middle two scores if there is an even number of scores. The left-hand column of Table 17-5 shows the calculation of the median for the creativity test. Since there were 16 scores, the median is halfway between the eighth score (6) and the ninth score, which is also 6.

The median can also be computed from a frequency distribution. (This is useful when there are too many scores to list individually.) The total number of cases in the *f* column is divided by 2. Counting either from the top or bottom of the distribution of the *f* column, the middle score is found, or halfway between the two middle scores if there is an even number of cases. This halfway point is the median. Calculation of the median from a frequency distribution is shown in the right-hand column of Table 17-5.

TABLE 17-5. Calculation of the Median Number of Uses for a Top Hat

Rank all scores from highest to lowest and find the middle score		*Find the score (X) of the middle case in the frequency distribution*		
Creativity score (X)		*X*	*f*	
11		11	1	
9		10	0	
8	N = 16 8 scores (50% of all scores) above	9	1	8th and 9th (middle cases fall here.
8		8	2	
7		7	3	
7		6	4	← Median = 6
7		5	3	
6		4	2	
Median = 6			*N* = 16	
6		One-half of 16 is 8. Find the 8th and 9th cases in the frequency distribution, counting either from the top or bottom. The median is the score lying between them.		
6				
6	8 scores (50% of all scores) below			
5				
5				
5				
4				
4				

Mode

The *mode* is the single score that occurs most often in a distribution. Of the three measures of central tendency, it is the easiest to compute but often the least informative, since it provides only a rough estimate of central tendency. Table 17-5 indicates that four people gave six uses for a top hat—more than gave any other single number of uses. The *mode* of this distribution is therefore 6. It is not unusual for a distribution to have more than one mode. When there are two modes, the distribution is called *bimodal*.

Measures of Variability

Measures of variability indicate the amount of spread or dispersion within the distribution of scores. In some studies, the scores show very little spread, clustering around the indicators of central tendency. In other studies, scores are greatly spread out. This information is important in the interpretation of research findings. For example, it can show whether or not the group average has been unduly influenced by a few extreme scores. The most commonly used measures of variability are the *range* and the *standard deviation*. The range provides a quick and easy assessment of spread. To compute the range, simply subtract the lowest score from the highest. On the creativity test, the highest score was 11 and the lowest was 4. The range is $11 - 4 = 7$.

The standard deviation is a much more complicated statistic, but it is also more informative and useful than the range, since it is included in many other statistical tests. The larger the standard deviation, the greater the spread of scores within the range. The standard deviation takes into account the distance of each score from the mean of the distribution.

A small standard deviation indicates that most scores are grouped around the mean. A larger standard deviation indicates greater variability of the scores from the mean (see Figure 17-4).

A formula for computing the standard deviation is:

$$\text{S.D.} = \sqrt{\frac{\Sigma X^2 - \frac{(\Sigma X)^2}{N}}{N}}$$

where

S.D. = standard deviation
ΣX^2 = each score squared and then summed
$(\Sigma X)^2$ = all scores summed and then squared
N = number of cases

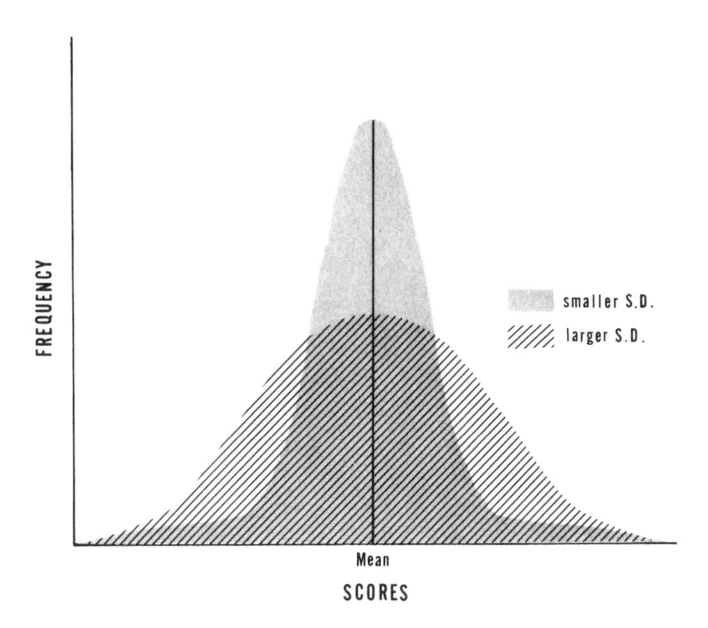

FIG. 17-4. Two curves with identical means and ranges but different standard deviations.

Do not be surprised if you find other formulas for calculating the standard deviation. They vary among statistics texts. Table 17-6 shows the calculation of the standard deviation for the results of the creativity study. In calculating a standard deviation, it is not necessary to list the scores from highest tc lowest. They can be written in any order. They are presented in order here to follow the previous example.

What is the purpose of this procedure? First, the standard deviation provides an indication of variability. Second, it is also used in many other statistical tests.

The reader may welcome a brief illustration of the practical significance of a standard deviation. If half the children in the arithmetic example used earlier had scored very high on the test and the other half had scored very low, the mean (average score) would have meant little. It would tell us nothing about the typical performance, because none of the children would have received an average score. Such a distribution (with cases spread far away from the mean) would show a high standard deviation. When presenting

TABLE 17-6. Calculation of the Standard Deviation for the Creativity Scores

Number of uses for a top hat X	X^2
11	121
9	81
8	64
8	64
7	49
7	49
7	49
6	36
6	36
6	36
6	36
5	25
5	25
5	25
4	16
4	16

$$\text{S.D.} = \sqrt{\frac{\Sigma X^2 - \frac{(\Sigma X)^2}{N}}{N}}$$

$$= \sqrt{\frac{728 - \frac{10{,}816}{16}}{16}}$$

$$= \sqrt{\frac{728 - 676}{16}}$$

$$= \sqrt{\frac{52}{16}}$$

$$= \sqrt{3.25}$$

$$\text{S.D.} = 1.80$$

$\Sigma X = 104 \quad \Sigma X^2 = 728 \quad N = 16$

$(\Sigma X)^2 = (104)^2 = 10{,}816$

research results, it is helpful to the reader to include, besides the mean, an index of variability such as the standard deviation.

Categorical Measures

As mentioned earlier, the descriptive statistics mentioned thus far apply primarily to continuous measures. However, frequencies and the mode are appropriate to calculate for categorical measures. For example, if a sample of voters is asked whom they prefer as president, the number of people preferring each candidate constitutes a frequency, and the most frequently made selection is the mode. It would be impossible to calculate a mean, median, or standard deviation.

Graphic presentation of categorical data is best made with bar graphs. Frequency curves should *not* be used with categorical data. They are misleading because the connecting lines suggest the existence of intermediate points that do not occur with distinct categories.

Summary

The original scores on some measure constitute the raw data. They are continuous if they fall along a single dimension. Frequencies refer to the number of times each score occurs. A percentile is the score at or below which a given percentage of the cases lie. The percentile rank of a score is the percentage of cases falling at or below that score. Bar graphs, histograms, and frequency polygons are figures used to illustrate distributions. Three measures of central tendency are the mean, the arithmetic average; the median, the midpoint of a frequency distribution; and the mode, the most frequently occurring score. The spread or variability among scores is indicated by the range and the standard deviation.

Further Reading

Bruning, J. L., & Kintz, B. L. (1987). *Computational handbook of statistics* (3rd ed.). Glenview, IL: Scott, Foresman.

Runyon, R. P., & Haber, A. (1988). *Fundamentals of behavioral statistics* (6th ed.). Reading, MA: Addison-Wesley.

Tufte, E. R. (1983). *The visual display of quantitative information.* Cheshire, CT: Graphics Press.

Zeisel, H. (1985). *Say it with figures* (rev. ed.). New York: Harper & Row.

Sample Problems

1. Here is the frequency distribution of running speed scores from several sixth-grade classes.

Time (in seconds)	*Frequency*
40–44	1
45–49	6
50–54	10
55–59	22
60–64	25
65–69	16
70–74	8
75–79	4
80–84	2

Use these scores to construct a bar graph or histogram. Be sure to label both the vertical and horizontal axes properly.

Calculate the cumulative frequencies percentage at each interval.

2. Here are the scores from 10 sixth-grade boys who participated in a study of running speed.

Name	*Time (in seconds)*
John H.	52
Henry B.	62
Alan C.	47
Henry C.	59
Tom R.	50
Jerry W.	72
Matt S.	50
Andy F.	71
Ken P.	48
Eric A.	50

Compute each of the following descriptive statistics for the above scores: mean*, median, range, standard deviation.*
*Use of a calculator will be helpful.

Answers to Problems

1. Cumulative frequencies: 2, 6, 14, 30, 55, 77, 87, 93, 94
   Cumulative percent: 2.1, 6.4, 14.9, 31.9, 58.5, 81.9, 92.6, 98.9, 100
2. Mean = 56.1
   Median = 51 (midpoint between 50 and 52)
   Range = 72 − 47 = 25
   Standard Deviation = 8.916*

*Some calculators use a slightly different formula, employing $N - 1$ rather than N in the denominator. In that case, the standard deviation is 9.398.

18 Inferential Statistics

"Smoke Gets in Your Eyes" is a very romantic song but not a very pleasant experience for many people. Kathryn Anthony (1977) wanted to find out whether people stood farther away from smokers than from nonsmokers. She recruited student volunteers from psychology classes. When they arrived at the designated room, each volunteer was introduced to the research associate (a person working with Ms. Anthony) and asked to "choose the distance most comfortable for conversation between you and this person, assuming you had just met." Once the student had created a comfortable conversational distance, a tape measure was used to calculate it in centimeters. During approximately half of the 76 sessions, the research associate was smoking a lit cigarette. During the other sessions, the associate did not smoke.

Anthony found that students stood a mean distance of 62.5 cm from the nonsmoking associate. When the associate was smoking, the distance was 80.3 cm.

From these average distances, it appeared that the cigarette had an effect on conversational distance. However, there may have been a large overlap in the scores, and perhaps only one or two extreme scores was responsible for these differences. For example, a single student who remained 4 m away from a smoking person might have been responsible for this entire difference. The problem involves deciding when a difference is reliable and when it is due to chance fluctuations, perhaps the result of a few extreme scores in one group. One solution is to repeat the study and see if similar differences are obtained. Another approach is to test the reliability of differences between groups using an *inferential* test. *Inferential statistics* permit generalization from samples to populations.

Chapter 16 on sampling described the process of generalizing from a *sample* to a *population,* making inferences from a subset to the whole. There were two problems that arose in doing this. One was *sampling error,* chance fluctuation that is inevitable when taking small samples from large populations. The second problem was *sample bias,* which comes from faulty sample selection resulting in a subset that does not accurately represent the population. Statistical techniques are used to take care of the problem of *sampling error,* the chance fluctuation one must expect. Statistical manipulation *cannot* solve problems of inadequate or biased sampling. That is a research design problem. Statistics cannot salvage an inadequately designed or poorly conducted piece of research. To repeat, statistics are used to deal with sampling error, to estimate the role of chance in the findings.

Note to the reader: If you have not taken a course in statistics, this chapter will seem more difficult than others because of the many new terms and formulas. It is probably best to read this chapter for understanding rather than to attempt to grasp all the details of each statistical test. Do not be concerned if you cannot understand immediately how these tests can be applied. This will come in time as you work with actual data.

Testing the Null Hypothesis

The idea that any differences in an experiment are due to chance fluctuations and that the independent variable has no effect on the dependent variable is called the *null hypothesis.* This is a conservative way of phrasing the research question; the investigator predicts that there will be no reliable difference between groups. In the preceding example, the null hypothesis is that the two samples tested (those approaching the smoker or the nonsmoker) do not differ in average conversational distance. In a correlational study, where one is looking at the relationship (or lack of it) between variables, the null hypothesis is that there is no relationship between them.

Following a statistical test of the null hypothesis, it is either accepted (no difference) or rejected (there is a reliable difference). If we find the differences could be attributed to chance, then we believe (accept) the null hypothesis. When the null hypothesis is accepted, the researcher concludes that the independent variable had no effect on the outcome. If the null hypothesis is *rejected* (the difference was *not* due to chance), then the conclusion is that there was an effect. Acceptance or rejection of the null hypothesis can be done quickly and easily when you know the techniques. Formulas and tables necessary for interpreting the results have been worked out by statisticians.

Alternative Hypothesis

The working hypothesis is stated in positive terms—that there *is* a difference between groups or that the independent variable will have a significant effect on the dependent variable. This is called the *alternative hypothesis* because it is an alternative to the null hypothesis.

Testing the null hypothesis feels odd because you are usually more interested in confirming your working (alternative) hypothesis. This is probably what you were interested in initially. Nevertheless, statistical practice requires that the fate of the alternative hypothesis actually rests on acceptance or rejection of the null hypothesis.

Probability Levels

In using inferential statistics, it is the null hypothesis that is tested. Whether or not your own hypothesis can be accepted depends on whether you are able to reject the null hypothesis that no reliable difference exists. Rejecting the null hypothesis is a gamble. The *probability level* of a statistic indicates the odds. If the probability of the difference occurring by chance alone is less than 5 in 100, than most researchers are willing to gamble that the outcome is *not* one of these five cases, and will therefore reject the null hypothesis and accept the research hypothesis that a genuine difference exists.

A common practice in the social sciences is to use this .05 probability level for testing the null hypothesis. This is also called the *.05 level of significance* and is abbreviated as $p < .05$. When the odds of a difference being due to chance are less than 5 in 100 ($p < .05$), the null hypothesis will be rejected. When the likelihood of a difference being due to chance is more than 5 in 100, then the null hypothesis—that the difference is probably due to chance—can be accepted. Some researchers prefer the .01 level of significance ($p < .01$), rejecting the null hypothesis only if the odds of a difference being due to random fluctuation are less than 1 in 100. The .01 probability level reduces the likelihood of mistaking a chance difference for a genuine difference.

Do not confuse statistical significance with practical importance. To say that a difference is "statistically significant" means only that it is not due to chance. With very large samples, even a tiny difference can be statistically significant even though it probably will not have much practical relevance.

Statistical Tests for Continuous Data

The two most common tests for continuous data are the *t* test and analysis of variance. The *t test* is used for comparing two groups. *Analysis of variance* is used for comparing differences among two or more groups. These tests take into account the amount of overlap in the distribution of scores in order to make a judgment about the significance of the *difference between means.* Before either test can be employed, there is an important condition that must be met: The distribution of scores in the population from which scores were drawn should approximately follow the normal curve. A perfect normal curve is the symmetrical bell-shaped curve shown in Figure 18-1. Characteristics measured in nature among large populations often produce this kind of frequency curve. If we measure the length of tigers' tails, the age at which children lose their milk teeth, or the production of fruit during the growing season, we would get a heavy concentration of scores in the middle with scores diminishing in frequency to the two ends of the curve. In a perfectly normal distribution, the mean, median, and mode are identical.

Most distributions of behavioral data do not perfectly match a bell-shaped curve but rather approximate it, as in Figure 18-2. This is acceptable for employing one of the tests for continuous data.

Sometimes a distribution of scores deviates markedly from a normal curve. This happens when there are many extreme scores at the upper or lower ends of the distribution, as in Figure 18-3. Such distributions are said to be *skewed.* It is necessary to know whether data approximate a normal curve or instead are skewed in order to select the best statistical technique for describing them and testing any differences between groups. When the

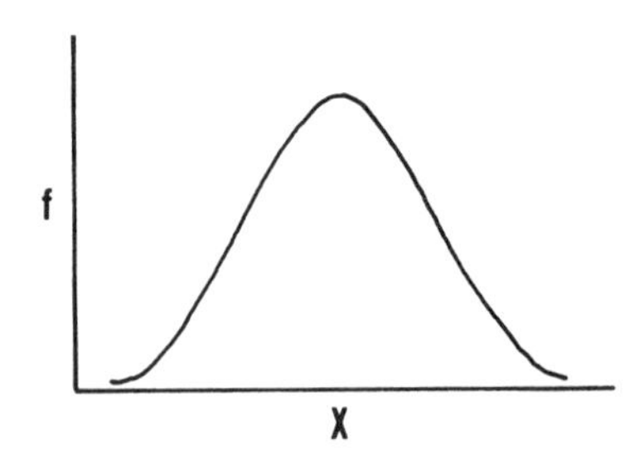

FIG. 18-1. Normal curve.

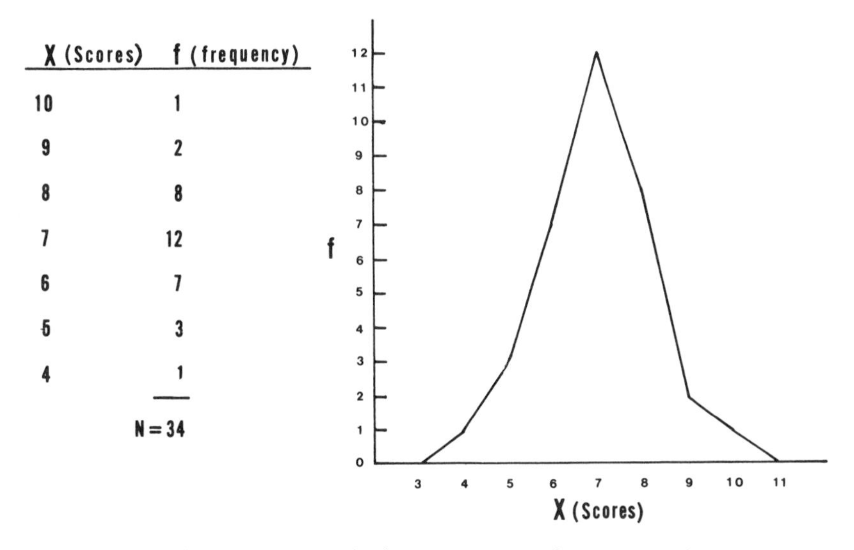

X (Scores)	f (frequency)
10	1
9	2
8	8
7	12
6	7
5	3
4	1
	N=34

FIG. 18-2. Normally distributed data and frequency polygon.

data are continuous and the frequency distribution is close to normal (it will rarely be a perfectly normal curve), then a *t* test or analysis of variance can be employed. If your results are skewed, but you think the characteristic being measured is probably normally distributed in the population, then you can still use a *t* test or analysis of variance. If you think the population data do not approximate a normal curve, then a statistical test for categorical data described later in this chapter should be used.

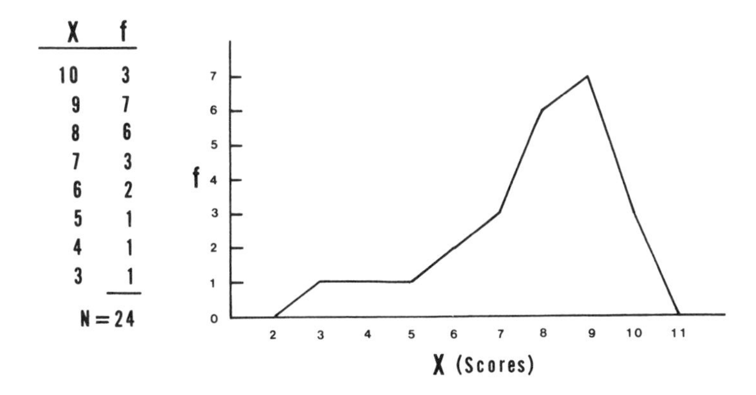

X	f
10	3
9	7
8	6
7	3
6	2
5	1
4	1
3	1
	N=24

FIG. 18-3. Skewed data and frequency polygon.

How to Do a t *Test*

The t test assesses the significance of the difference between the means of two groups of scores. The numerical result of this test is called a *t ratio.* There are two formulas for a t test. One is used when the two groups are independent; the other is used when the two groups compared have been matched on some characteristic or the same people have been tested before and after an experimental treatment. The latter are called *matched scores* or *correlated measures.*

The formula for a t test for independent groups and an example of its use are shown in Box 18-1.

BOX 18-1. Calculation of a t test for Two Independent Groups

Two groups of employees are given a test. One group contains five people, the other six. Do the employees in Group 1 score significantly higher than the employees in Group 2, or is the difference between them the result of random fluctuation?

Scores on Test	
Group 1 (X)	*Group 2 (Y)*
4	4
5	9
5	2
3	3
3	3
	4

The formula for a t test for independent groups is

$$t = \frac{\overline{X} - \overline{Y}}{\sqrt{\left[\dfrac{\left(\Sigma X^2 - \dfrac{(\Sigma x)^2}{N_x}\right) + \left(\Sigma Y^2 - \dfrac{(\Sigma y)^2}{N_y}\right)}{N_x + N_y - 2}\right] \cdot \left[\dfrac{1}{N_x} + \dfrac{1}{N_y}\right]}}$$

X = score in Group 1
Y = score in Group 2
N_x = number of scores in Group 1
N_y = number of scores in Group 2

Computation

Step 1. Add up scores in Group 1: $\Sigma X = 20$.
2. Square total for all scores in Group 1: $(\Sigma X)^2 = 400$.
3. Square *each score* in Group 1. Add together $\Sigma X^2 = 84$.

4. Find number of scores in Group 1: $N_x = 5$.
5. Compute mean for Group 1: $\overline{X} = \dfrac{\Sigma X}{N} = 4.0$.

(Steps 6–10. Repeat steps for Group 2)

6. $\Sigma Y = 25$.
7. $(\Sigma Y)^2 = 625$.
8. $\Sigma Y^2 = 135$.
9. $N_y = 6$.
10. $\overline{Y} = 4.17$.
11. Enter these figures in the formula for the t test for independent groups.

$$t = \frac{4.0 - 4.17}{\sqrt{\left[\dfrac{\left(84 - \dfrac{400}{5}\right) + \left(135 - \dfrac{625}{6}\right)}{5 + 6 - 2}\right] \cdot \left[\dfrac{1}{5} + \dfrac{1}{6}\right]}}$$

$$= \frac{-.17}{\sqrt{\left(\dfrac{4 + 30.83}{9}\right)(.20 + .17)}} = \frac{-.17}{\sqrt{(3.87)(.37)}}$$

$$= \frac{-.17}{\sqrt{1.43}} = \frac{-.17}{1.20} = -.14$$

12. Compute degrees of freedom $df = (N_x - 1) + (N_y - 1) = 4 + 5 = 9$. Check Table A-1 in Appendix A. Look at 9 degrees of freedom. Table A-1 shows that a t ratio of 2.26 or greater is required for significance at the .05 level with 9 degrees of freedom. Since $t = .14$ (the minus sign can be ignored), this difference is *not* significant.

The t Test for Matched Scores

When two sets of scores have been matched in some way, such as the same individuals tested before and after an experimental treatment, the t test for matched scores is used. (See Box 18-2.)

Interpreting the t *Ratio*

Having computed the t ratio according to one of the formulas shown in the boxes, the next step is to determine the *degrees of freedom* (*df*). This is a fairly complicated concept to understand. Fortunately, it is very easy to compute. For a t test with independent scores, the degrees of freedom are

BOX 18-2. Calculation of a t Test for Matched Scores

A group of workers is given a hand steadiness test before and after taking a drug. Does the drug have a significant effect on their performance?

Score on Hand Steadiness Test				
Employee name	*Before*	*After*	d	d^2
Bill	72	70	+2	4
John	81	78	+3	9
Mary	75	74	+1	1
Sue	77	78	−1	1
Al	84	80	+4	16
Sam	72	72	0	0

Step 1. Compute the difference (d) between each person's scores before and after taking the drug. Subtract each *after* score from each *before* score. Pay close attention to the sign (plus or minus) of each difference (d).

2. Square each difference (d^2).
3. Compute $\Sigma d = 9$.
   Note: Pay close attention to plus and minus signs.
4. Compute $(\Sigma d)^2 = 81$.
5. Compute $\Sigma d^2 = 31$.
   Note: All numbers here are positive since they have been squared.
6. $N = 6$.
7. Enter figures into formula. The formula for a t test for matched scores is:

$$t = \frac{\dfrac{\Sigma d}{N}}{\sqrt{\dfrac{\Sigma d^2 - \dfrac{(\Sigma d)^2}{N}}{N(N-1)}}}$$

d = difference between the matched scores
N = number of *pairs* of scores

Computation

$$t = \frac{\dfrac{\Sigma d}{N}}{\sqrt{\dfrac{\Sigma d^2 - \dfrac{(\Sigma d)^2}{N}}{N(N-1)}}} = \frac{\dfrac{9}{6}}{\sqrt{\dfrac{31 - \dfrac{81}{6}}{6(6-1)}}}$$

$$= \frac{1.5}{\sqrt{\dfrac{31 - 13.5}{30}}} = \frac{1.5}{\sqrt{.58}} = \frac{1.5}{.76} = 1.97$$

8. Compute degrees of freedom. For related scores, $df = N - 1 = 6 - 1 = 5$.
9. Check Table A-1 in Appendix A. The critical value for t to be significant at the .05 level for 5 degrees of freedom is 2.57 or greater. Since this t ratio is only 1.97, it is *not* significant.

the total number of scores less one for each group. For a t test for matched scores, the degrees of freedom are the number of matched pairs less one.

The final step in interpretation is to consult Table A-1 in Appendix A. The table is used to decide whether an obtained t ratio is due to chance. Using the .05 level of statistical significance, a probability of *less than* 5 in 100 that an obtained difference is due to chance calls for *rejection* of the null hypothesis (i.e., concluding that the obtained difference is reliable). When the probability that an obtained difference is due to chance is *more than* 5 in 100, the null hypothesis (no reliable difference) is *accepted.* If the investigator wants to be very certain of the results, a probability level of 1 in 100 (.01) may be used.

To find the significance of a t ratio in Table A-1, first consult the left column and find the degrees of freedom closest to those in your study.

Example: In the cigarette study there were two unrelated groups of 39 and 37 subjects. This produced 74 degrees of freedom ($39 - 1 + 37 - 1$). Since this table does not show $df = 74$, the nearest degrees of freedom (100) are used. Now move to the right of $df = 100$ to find the appropriate t ratios for the .05 and .01 probability levels. These are 1.98 and 2.63, respectively. Since the obtained t ratio between the groups (3.36) is larger than either of these figures, we conclude that the difference is reliable at the .01 level of significance, and the null hypothesis can be rejected.

Analysis of Variance (ANOVA)

Analysis of variance, sometimes abbreviated ANOVA, can be used when there are more than two sets of continuous scores. The principles of interpreting differences are similar to those for the t test. Like the t test, ANOVA is used to judge the reliability of differences between means. Its advantage is that it can be used for more than two means at a time or it can be used in place of t for two means. An additional benefit is that ANOVA can also be used to assess the effects of more than one independent variable at a time. However, our example will involve an experiment with only one independent variable.

One-way ANOVA

An Air Force researcher is interested in the effect of sleep deprivation on information processing by air traffic controllers. The independent variable is sleep deprivation measured at three levels: none, slight, and moderate deprivation. The dependent variable is the number of mistakes made on the task. Four subjects are randomly assigned to each group, giving a total of 12 individuals in the study. The results are in Table 18-1.

The question to be answered by the statistical analysis is whether or not the means of these three groups differ significantly from one another (i.e.,

TABLE 18-1. Sleep Deprivation Data

Group 1		*Group 2*		*Group 3*	
No deprivation	*Number of mistakes*	*Slight deprivation*	*Number of mistakes*	*Moderate deprivation*	*Number of mistakes*
Subject 1	1	Subject 5	3	Subject 9	2
Subject 2	2	Subject 6	2	Subject 10	3
Subject 3	0	Subject 7	3	Subject 11	4
Subject 4	2	Subject 8	4	Subject 12	4
Means	1.25		3.0		3.25

more than to be expected by chance). We are concerned with the differences *between* groups, but we gauge the chance variation by looking at the variability of scores *within* each group. In this example we will get a measure of the differences between the groups, with respect to the number of mistakes, and judge that against the distribution or spread of mistakes within each group. The between group differences must be significantly greater than those within groups in order for the differences between means to reach statistical significance.

Looking at the means in this example, it appears that the sleep deprivation leads to more errors, but the differences between them are not large and could reasonably be attributed to chance. The ANOVA will allow us to decide the odds of that being the case. The statistical calculation requires computing *sums of squares (SS),* which take into account the spread of scores within the sample. The calculations are entered into an ANOVA table, which is shown at the end of Box 18-3. Going through the necessary steps, the F ratio is generated. It is a ratio of the average between-group to the average within-group variability. Its statistical significance is evaluated using Table A-6 in the Appendix.

Additional tests called *mean comparisons* can be made to find exactly where the significant difference lies. In our example, the difference between groups 1 and 2, and between 1 and 3 was significant, while the difference between groups 2 and 3 was not.

There are many more variations of the ANOVA technique that extend its range of applications.

ANOVA for Groups of Unequal Size

In natural experiments or observational studies, it is common to have unequal numbers of subjects at various levels of the independent variable. When this occurs, the calculations must be adjusted to take into account the unequal numbers. The underlying assumptions of ANOVA are such that it is preferable to have equal numbers in each group whenever possible. For

BOX 18-3. Calculation of a One-Way Analysis of Variance (ANOVA)

Step 1. Arrange the scores by groups.

Group 1		*Group 2*		*Group 3*	
No deprivation	*Number of mistakes*	*Slight deprivation*	*Number of mistakes*	*Moderate deprivation*	*Number of mistakes*
Subject 1	1	Subject 5	3	Subject 9	2
Subject 2	2	Subject 6	2	Subject 10	3
Subject 3	0	Subject 7	3	Subject 11	4
Subject 4	2	Subject 8	4	Subject 12	4

The results of each set of steps are entered on an ANOVA table:

ANOVA Table

Source	SS	*df*	MS	*F*
Total	SS_{tot} ______			
Treatment (deprivation groups)	SS_{bet} ______			
Error	SS_{within} ______			

A. Calculation of the total sum of squares (SS_{tot}).
NOTE: If you are using a calculator with a X^2 key, you can do steps 2 and 3 at the same time.

Step 2. Add the scores in *each* group. This is the *sum* for each group.

	Group 1	*Group 2*	*Group 3*
	1	3	2
	2	2	3
	0	3	4
	2	4	4
Sum	5	12	13

Step 3. Square each individual score and add these squared values together.
$1^2 + 2^2 + 0^2 + 2^2 + 3^2 + 2^2 + 3^2 + 4^2 + 2^2 + 3^2 + 4^2 + 4^2$
$= 1 + 4 + 0 + 4 + 9 + 4 + 9 + 16 + 4 + 9 + 16 + 16$
$= 92$ This is ΣX^2

Step 4. Add all the group sums (in step 2) together to get the sum for all scores, the grand total.
$5 + 12 + 13 = \Sigma X_{tot} = 30$

Step 5. Square the grand total (step 4).
$(\Sigma X_{tot})^2 = (30)^2 = 900$

Step 6. Divide the result of step 5 by N, the total number of cases.

$N = 4 + 4 + 4 = 12$

$$\frac{(\Sigma X_{tot})^2}{N} = \frac{900}{12} = 75$$

This is called the *correction factor.*

Step 7. Subtract the *correction factor* (step 6) from the result of step 3, the sum of the squared values.

$\Sigma X^2 -$ (correction factor) $= 92 - 75 = 17.0$

This is the *total sum of squares* (SS_{tot}).

Enter it in the ANOVA table (see model table in Section H).

B. Calculation of the between group sum of squares (SS_{bet}).

Step 8. For *each* group, square the sum (step 2) and divide by the number of scores in the group.

$$\frac{5^2}{4} + \frac{12^2}{4} + \frac{13^2}{4} = 84.5$$

Step 9. Subtract the correction factor (step 6) from the value in step 8.

$84.5 - 75 = 9.5$

This is the between groups sum of squares (SS_{bet}).

Enter this value in the ANOVA table.

C. Calculation of the within group sum of squares (SS_{within}).

Step 10. Subtract the between group sum of squares (step 9) from the total sum of squares (step 7)

$17 - 9.5 = 7.5$

This gives the within group sum of squares (SS_{within}) Enter in the ANOVA table.

D. Calculate degree of freedom (df) and enter on the ANOVA table.

Step 11. For SS_{tot} $\quad df_{tot} =$ (total number of cases) $- 1$

$df_{tot} = 12 - 1 = 11$

For SS_{bet} $\quad df_{bet} =$ (number of groups) $- 1$

$df_{bet} = 3 - 1 = 2$

For SS_{within} $\quad df_{within} = df_{tot} - df_{bet}$

$df_{within} = 11 - 2 = 9$

E. Calculate mean squares (MS) for the *between* and *within* group measures.

Step 12. The mean squares (MS) are computed as SS/df. (Note: MS_{tot} is not needed.)

$$MS_{bet} = \frac{SS_{bet}}{df_{bet}} = \frac{9.5}{2} = 4.75$$

$$MS_{within} = \frac{SS_{within}}{df_{within}} = \frac{7.5}{9} = .83$$

Enter these in the ANOVA table.

F. Calculate the F ratio.

Step 13.

$$F = \frac{MS_{bet}}{MS_{within}} = \frac{4.75}{.83} = 5.72$$

Enter the value on the ANOVA table.

G. Statistical significance.

Step 14. See Table A-6 in Appendix A to find the probability associated with the F ratio.

There were 2 degrees of freedom (df) in the numerator of the F ratio, and 9 df in the denominator. Using Table A-6, first go across the columns until you reach 2 df for the numerator, then proceed down this column until you reach 9 df for the denominator. This shows that for the differences among the means to be significant at the .05 level, the obtained F ratio must equal or exceed 4.26, and to be significant at the .01 level, must equal or exceed 8.02.

Because the obtained statistic, 5.72, exceeds the critical value at the .05 level, we can reject the null hypothesis at the .05 level, and conclude that sleep deprivation has a negative impact on task performance. See the notation, $p < .05$, on the ANOVA table.

H. Model ANOVA Table.

Source	SS		df	MS	F
Total	SS_{tot}	17.0	11		
Treatment (deprivation groups)	SS_{bet}	9.5	2	4.75	5.72 $p < .05$
Error	SS_{within}	7.5	9	0.83	

this reason, some researchers will use a random number procedure to eliminate scores from the larger groups to bring all groups to an equal size. There are formulas available in statistics books for ANOVA with groups of unequal size. Most computer programs can handle unequal sample sizes.

Factorial Design

The example in Box 18-3 illustrates the steps for doing a one-way ANOVA. It is called one-way because it involves only one independent variable or factor. When more than one factor or independent variable is introduced, the design is termed *factorial* (i.e., two-factor or two-way ANOVA, three-factor or three-way ANOVA, etc.). When additional factors are introduced, interpretation of the ANOVA becomes more complex.

The research design for the sleep deprivation study could be modified by introducing three levels of job experience as an additional factor: less than 1 year experience, 1–5 years experience, and more than 5 years experience. Such a design would allow for (1) assessment of the effects of sleep deprivation, (2) assessment of the effects of job experience, and (3) an assessment of a possible *interaction* between the two. For example, sleep deprivation may be of less importance for experienced than for beginning air controllers.

The potential for judging interactions makes the factorial design a very useful one.

Repeated Measures Design

This design is used when the same subjects are studied at different times or under different conditions. In the air controller example, it would have been possible to study the same subjects under the different conditions of sleep deprivation.

The repeated measures design is very powerful or sensitive as it permits the removal of individual differences in gauging the effects of the independent variable. In the statistical calculation, subjects becomes a second factor and subject-by-treatment interaction serves as the error term for assessing chance variability.

Analysis of Covariance

This technique takes into account scores on some other variable believed to affect the action of the independent variable. Instead of introducing amount of experience as a second factor (independent variable) in our treatment of the groups, we could statistically control for it by adding years of experience for each subject into our calculations. It would not be manipulated as an independent variable, but its effect could be taken into account. Analysis of covariance in this example would tell us the effects of sleep deprivation, after making an adjustment for job experience.

Multivariate Designs

In ANOVA, only one dependent variable is used per analysis. There are also *multivariate* procedures in which the relationship between two or more dependent variables enters into the statistical calculation. These are complex and generally done by computer. While relatively easy to generate because the computer does the work, their interpretation requires a well-grounded understanding of statistical concepts and inference.

These are only a few of the important variations of the ANOVA technique. They are beyond the scope of an introductory course in research methods. Many colleges and universities teach advanced statistics courses devoted entirely to analysis of variance. Additional information on ANOVA can be found in the books listed at the conclusion of this chapter.

A Statistical Test for Categorical Data: Chi-Square

A university class in research methods undertook a series of observational studies. One study was designed to see whether men or women bicyclists were more likely to ignore a stop sign at a major intersection when leaving the college campus. The researchers thought that there might be a sex difference, but were not sure which direction it would take. They made 30-minute observations during weekday afternoons over a 2-week period. The observations produced categorical data. Each cyclist was classified as having stopped or not stopped at the sign. Since the *t* test could not be used with categorical data like these, the researchers chose to use a Chi-square test. [Chi (χ) is a Greek letter pronounced *ki,* as in *kite.*]

The Chi-square statistic is used to test whether the actual results (such as the bicycle observations) differ reliably from what could be expected by chance. The two key sets of figures in this test are the *obtained values* (*O*) and the *expected values* (*E*). The *obtained values* (*O*) are the raw data arranged in the appropriate cells of a *contingency table* with the totals in the margins (see Box 18-4). This arrangement is called a contingency table because being in cell A is contingent in this case on being male and having stopped at the sign.

The *expected values* (*E*) are those that would occur due to chance alone. The marginal totals, the characteristics of the sample as a whole, are used to calculate the *E* values for each cell in the contingency table. For example, in the bicycle study, out of the entire sample of people, 192 stopped at the sign, which represents 192/301 of the sample. Furthermore, 91 of the people in the sample were male. Therefore, based on the characteristics of the entire sample, if no other factors were influencing behavior, we would expect 192/301 of the 91 males to have stopped at the sign. This gives us our *E* value of 58.05. This is the rationale for the calculation of the expected values. Box 18-4 shows a more simple procedure for calculating the *E* values.

After the *E* values are calculated, the Chi-square statistic is computed. The formula for the Chi-square and a computational example using the bicycle observations are shown in Box 18-4.

Interpretation of the Chi-square follows the same general principles as were used in the *t* test. Table A-2 in Appendix A is used to determine whether the Chi-square is statistically significant at the .05 or .01 level. When the computed value is larger than the figures shown in the table under the appropriate degrees of freedom, we can conclude that the difference is reliable.

The Chi-square test may be used for any number of rows and columns. However, when there are more than four rows or four columns, interpretation of the results becomes awkward. There is no need to have the same number of rows and columns.

Example:
The Chi-square can be used with categorical data in each of the following forms.

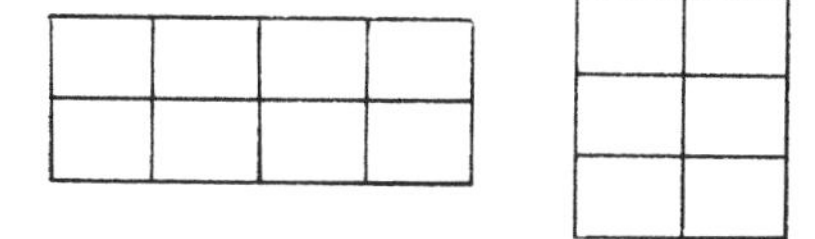

Contingency table patterns.

BOX 18-4. Calculation of Chi-square for Bicycle Observations

Step 1. Arrange the obtained data in the form of a *contingency table.* For the bicycle study, this is a table with two columns and two rows (2 × 2 table). Each cell shows the number of individuals meeting the conditions described in the margins—i.e., 66 males stopped at the sign. As a convention, the individual cells of the table are lettered.

Step 2. Add up the totals for each row ($A + B$, $C + D$) and for each column ($A + C$, $B + D$). Note: rows are horizontal, columns vertical.

		Response to stop sign		
		Stop	*No stop*	
Sex	Male	*A* 66	*B* 25	91 = 66 + 25
	Female	*C* 126	*D* 84	210
		66 + 126 = 192	109	301 = 91 + 210
				Check: 192 + 109 = 301

Step 3. Compute the expected frequencies (E). This is done by multiplying each column total by each row total and dividing by the grand total (301).

Cell A $\frac{192 \times 91}{301} = 58.05$

Cell B $\frac{109 \times 91}{301} = 32.95$

Cell C $\frac{192 \times 210}{301} = 133.95$

Cell D $\frac{109 \times 210}{301} = 76.05$

As a check, the values of E must also add up to the marginal totals.

Step 4. For each cell, subtract the expected value (E) from the observed value (O), square this figure, and divide by E. Sum the resulting numbers.

Formula: $\chi^2 = \Sigma \frac{(O - E)^2}{E}$

Cell A $\frac{(66 - 58.05)^2}{58.05} = \frac{(7.95)^2}{58.05} = 1.09$

Cell B $\frac{(25 - 32.95)^2}{32.95} = \frac{(7.95)^2}{32.95} = 1.92$

Cell C $\frac{(126 - 133.95)^2}{133.95} = \frac{(7.95)^2}{133.95} = .47$

Cell D $\frac{(84 - 76.05)^2}{76.05} = \frac{(7.95)^2}{76.05} = .83$

$\chi^2 = 1.09 + 1.92 + .47 + .83 = 4.31$

Step 5. Compute the degrees of freedom for the contingency table. Formula for degrees of freedom for the Chi-square:

df = (number of rows − 1)(number of columns − 1)

$df = (2 - 1)(2 - 1) = 1$

Step 6. To find the significance of the Chi-square, consult Table A-2 in Appendix A. Check the left-hand column under the appropriate degrees of freedom ($df = 1$) and then move to the right to the appropriate probability level. Table A-2 shows that with one degree of freedom, a Chi-square of 3.84 or greater is required to reject the null hypothesis at the .05 level. Since the Chi-square for the bicycle observations is larger than this, we can conclude that there was a reliable sex difference in stopping at this intersection. The direction of the difference can be seen in the contingency table used in Step 2. This shows that a higher percentage of men than women respected the stop sign at this intersection.

For a 2×2 contingency table (two rows and two columns), there is a direct formula for Chi-square that does not require you to calculate the expected frequencies. This is shown in Box 18-5.

The Chi-square becomes unreliable when the *expected* frequencies (E) are very small (i.e., less than 5). It should never be used when an expected frequency is 0. Note that this applies only to E rather than to the observed frequencies (O), which can be any size, including zero.

Note to the reader: This is another good place to stop and review this material.

BOX 18-5. Calculation of Chi-square from a 2 × 2 Contingency Table

Contingency table:

A	B	$A + B$
C	D	$C + D$
$A + C$	$B + D$	

$$\chi^2 = \frac{N(AD - BC)^2}{(A + B)(A + C)(B + D)(C + D)}$$

Using the data from the example in Box 18-4,

$$\chi^2 = \frac{301[(66 \times 84) - (25 \times 126)]^2}{(91)(192)(109)(210)} = \frac{301(2394)^2}{399{,}934{,}080}$$

$$= \frac{1{,}725{,}102{,}036}{399{,}934{,}080} = 4.31$$

Correlation

The statistical tests mentioned thus far have dealt with differences between groups either by testing differences between means (*t* test and ANOVA), or by comparing observed with expected frequencies of occurrence (Chi-square). Another issue of concern to the researcher is *correlation.* This refers to an association between two sets of scores. For example, is there a relationship between years in college and a person's income at middle age? One way to answer this question is to plot the data on a graph in the form of a scattergram. The vertical axis in Figure 18-4 shows income and the horizontal axis shows years in college.

The data from a group of 20 individuals have been plotted on this graph. The pattern of dots shows a fairly clear trend for income to rise with years in college. This is a *positive relationship,* since an increase in one variable is accompanied by an increase in the other. It is also possible for two variables to be *negatively correlated;* that is, as one variable increases, the other variable decreases. This is also called an *inverse relationship.* An example of a negative correlation would be that between midsummer temperature and physical activity. As the days become hotter, activity declines. Figure 18-5 shows scattergrams for a positive relationship, a negative relationship, and no relationship between two variables.

A scattergram provides an instant visual picture of the association between two variables. For many purposes, a more precise indication of the degree of relationship is necessary. This calls for the use of a *correlation coefficient.* A correlation coefficient of +1.0 indicates a perfect positive relationship like that shown in the first scattergram in Figure 18-5. A perfect nega-

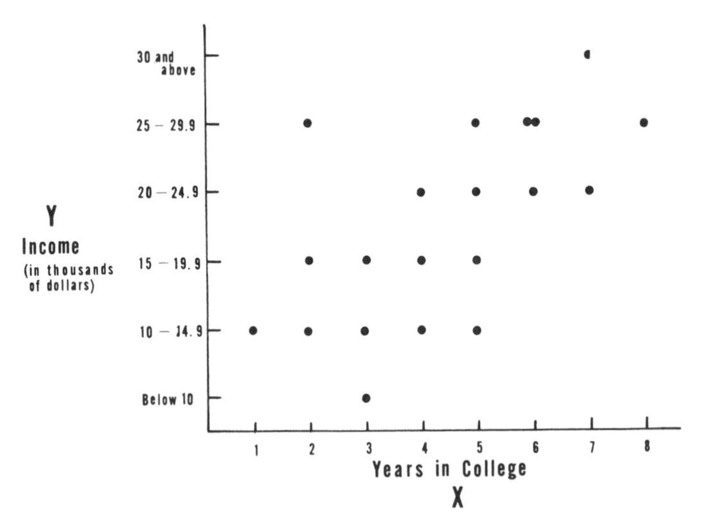

FIG. 18-4. Scattergram of the relationship between years in college and income in middle age (N = 20).

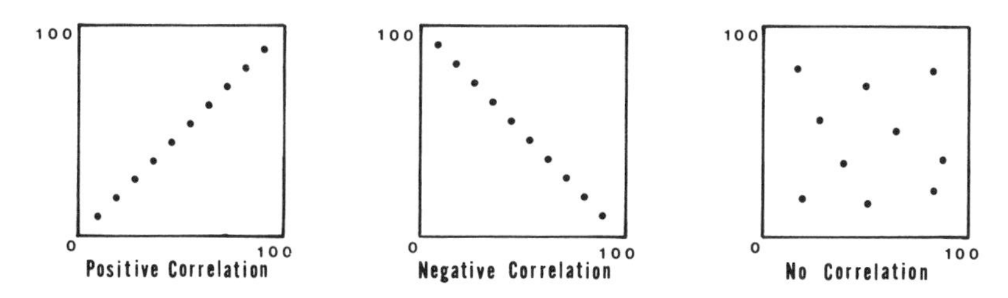

FIG. 18-5. Scattergrams produced by positive, negative, and absence of correlation.

tive relationship, such as that shown in the second scattergram, produces a correlation coefficient of −1.0. The absence of any relationship, such as in the third scattergram, produces a correlation coefficient of 0. The higher the correlation coefficient, either in a positive or a negative direction, the stronger the association between two variables.

The correlation coefficient for the scattergram in Figure 18-4 is .62, showing a strong relationship between education and income. Another example of a positive correlation is .87 between IQ test scores when people are tested twice, demonstrating strong test–retest reliability. In contrast, a correlation coefficient of .02 between intelligence and musical ability is extremely low and has no predictive value.

Correlation does not prove causation. The fact that two variables are correlated does not mean that one causes the other. In the example of college education and income, it is tempting to assume that going to college produced the higher income during middle age. However, it is likely that people who attended college for several years came from wealthier families in the first place. It is possible that some factor *other* than the two variables being measured may be the cause of the relationship.

The Correlation of Continuous Data: Pearson Product-Moment Coefficient (r)

When both sets of scores to be correlated are continuous, a Pearson product-moment coefficient (r) is a widely used statistic. The null hypothesis for a correlation states that there is no relationship between two sets of scores. If the probability level of the coefficient is less than .05 (or .01, to be more conservative), we can reject the null hypothesis and conclude that the two variables are correlated. The formula and a computational example for r are shown in Box 18-6.

Correlating Ranked Data: Spearman Rank-order Coefficient (rho)

The Pearson coefficient (r) is strongly influenced by extreme scores. If one person in the previous example had an income of $1 million a year, perhaps

BOX 18-6. Computation of the Pearson Product-Moment Coefficient (r)

A researcher tested the relationship between arithmetic and spelling test scores among a group of sixth graders. The hypothesis tested was that students who do well in arithmetic are also good spellers (a positive correlation). The highest possible arithmetic score was 10, and the highest possible spelling score was 20.

Student	*Arithmetic score* X	*Spelling score* Y	X^2	Y^2	XY
Jeanine	8	13	64	169	104
Howard	6	8	36	64	48
Tamara	5	7	25	49	35
Blossom	10	14	100	196	140
Hugh	7	12	49	144	84
Bradford	9	15	81	225	135
Clayton	2	3	4	9	6
Lorraine	3	3	9	9	9

$N = 8 \quad \Sigma X = 50 \quad \Sigma Y = 75 \quad \Sigma X^2 = 368 \quad \Sigma Y^2 = 865 \quad \Sigma XY = 561$
$(\Sigma X)^2 = (50)^2 = 2500$
$(\Sigma Y)^2 = (75)^2 = 5625$

Computation

Step 1. Sum X scores: $\Sigma X = 50$.
Step 2. Square ΣX: $(\Sigma X)^2 = 2500$.
Step 3. Square each X score. Add together $\Sigma X^2 = 368$.
(Steps 4–6 repeat with Y scores)
Step 4. $\Sigma Y = 75$.
Step 5. $(\Sigma Y)^2 = 5625$.
Step 6. $\Sigma Y^2 = 865$.
Step 7. Multiply each X score by its corresponding Y score. Sum all XY products. $\Sigma XY = 561$.
Step 8. Find number of pairs: $N = 8$.
Step 9. Insert these figures into the formula for r:

$$r = \frac{N\Sigma XY - (\Sigma X)(\Sigma Y)}{\sqrt{[N\Sigma X^2 - (\Sigma X)^2][N\Sigma Y^2 - (\Sigma Y)^2]}}$$

$$r = \frac{8(561) - (50)(75)}{\sqrt{[8(368) - 2500][8(865) - 5625]}} =$$

$$\frac{4488 - 3750}{\sqrt{(2944 - 2500)(6920 - 5625)}} = \frac{738}{\sqrt{(444)(1295)}} =$$

$$\frac{738}{\sqrt{574{,}980}} = \frac{738}{758.27} = .97$$

Step 10. Compute the degrees of freedom. The formula for degrees of freedom for r is: $df = N - 2$, where N is the number of pairs. $df = 8 - 2 = 6$.
Step 11. Interpreting the significance of r. Consult Table A-3 in Appendix A. First, look down the vertical column to $df = 6$; then move to the right, to the column indicating the .05 level of confidence. This table indicates that

when $df = 6$, a coefficient must be at least .707 to be statistically significant at the .05 level. Incidentally, this value of .707 holds for both positive and negative coefficients. The sign of the coefficient is unimportant in testing whether or not it is statistically significant. The sign *is* important in determining the direction of the relationship. Since the obtained coefficient in this example is .97, the relationship is reliable and the null hypothesis can be rejected. The sign of the coefficient (+) indicates that there is a positive association between the two variables; that is, as one increases, the other also increases.

from a trust fund, this would have seriously distorted the correlation in the direction of that single case. When the data are not normally distributed, one possibility is to use ranks instead of actual scores. The person with the highest income receives a rank of 1, the person with the next highest income receives a 2, and so on. This automatically reduces the influence of extreme scores. Two people with the same income would receive tied ranks. To rank tied scores, add together the ranks that would be occupied by these individuals and divide by the number of individuals obtaining the score.

Example:

Income	*Rank*
$50,000	1
$48,000	2
$46,000	3.5
$46,000	3.5
$45,000	5
$41,000	7
$41,000	7
$41,000	7
$40,000	9

The two individuals earning $46,000 occupied the third and fourth positions. Their ranks are computed by summing $3 + 4 = 7$ and dividing by 2 (number of individuals tied for that rank) = the tied rank score of 3.5. Remember that when you assign tied ranks to two individuals, you use up two ranks in the series. In the preceding example, the two individuals earning $46,000 both received ranks of 3.5, and the next person in the series receives a rank of 5. The three individuals earning $41,000 are tied for the sixth, seventh, and eighth places, respectively. To compute their tied rank scores, add all three ranks together and divide by the number of individuals ($6 + 7 + 8 = 21 \div 3 = 7$). Each of the three individuals earning $41,000 receives a tied rank score of 7 (a combination of ranks 6, 7, and 8). Therefore, the next individual receives a rank of 9.

When individuals are assigned a rank on each of the two measures, a Spearman rank-order coefficient *(rho)* is used to compute the correlation. The formula and computation of *rho* are shown in Box 18-7. The *rho* coefficient is most useful when there is a small number of cases. It would be very time consuming to rank more than 30 scores.

BOX 18-7. Computation of the Spearman Rank-order Coefficient *(rho)*: Rankings by the Team Coach of the Running Skill of Team Members Correlated with their Performance in a Track Meet

Runner	*Coach's ranking before race* X	*Order of finish in race* Y	d	d^2
MacClean	1	4	3	9
Smolenski	2	6	4	16
Jones	3.5	7	3.5	12.25
Jackson	3.5	1	2.5	6.25
Heinz	5	2	3	9
Baker	6	5	1	1
Harris	7	3	4	16

$\Sigma d^2 = 69.50$

Before beginning the computation, check to see that all ranking has been done correctly.

Computation

Step 1. Compute the difference (d) between the coach's rankings and the actual performance for each individual. You do not need to pay attention to plus or minus signs, since the numbers will be squared.

Step 2. Square each difference and sum them: $\Sigma d^2 = 69.50$.

Step 3. Find number of pairs: $N = 7$.

Step 4. Enter these figures in the formula: $rho = 1 - \frac{6\Sigma d^2}{N(N^2 - 1)}$

$$rho = 1 - \frac{6(69.50)}{7(49 - 1)} = 1 - \frac{417}{336} = 1 - 1.24 = -.24$$

Step 5. Evaluating *rho*. Table A-4 in Appendix A shows the significance levels for *rho*. First, look in the left-hand column under N. Note that this column uses the number of pairs (N) rather than degrees of freedom. This table shows that with an N of 7, a *rho* of .714 or greater (+ or −) is required for significance at the .05 level. Because the obtained coefficient in this example was −.24, we can accept the null hypothesis and conclude that there was no reliable relationship between the coach's ranking and the runner's performance in the race.

Correlating a Continuous Variable with a Ranked Variable

If one of the variables is continuous and the other is ranked, it is easy to transform the continuous scores into ranks and use *rho.* Simply rank the continuous scores from highest to lowest and then use the ranks to compute *rho.*

Other Measures of Correlation

Still other correlation coefficients are useful under special circumstances. When one variable has been artificially split into two categories (e.g., between people of age 30 and over and those under 30) and the other is in the form of a continuous distribution (e.g., a score on an attitude scale), the Biserial coefficient may be used. When both variables have been artificially split into two categories, the Tetrachoric coefficient can be used. Formulas and methods for computing these coefficients can be found in most statistical textbooks. However, while these coefficients are appropriate under special circumstances, the Pearson r is the most stable and reliable measure of correlation and should be used wherever possible.

Limitations of This Chapter

This chapter is intended only as an introduction to a few basic techniques. Of necessity, many other useful statistical tests have been omitted. The student who would like to know more about statistical analysis should consult a textbook in the area. There are books available for the beginner as well as for advanced students.

At this point, we will repeat something said at the beginning of this chapter: Do not be concerned if you have not understood everything about inferential statistics. The origin of many statistical tests goes far beyond what is needed to use them in practice. Once you start working with actual data, the meaning of the tests, as well as their limitations, will become more clear.

Summary

Inferential statistics are used to take sampling error (chance variation) into account in interpreting research results. They cannot eliminate problems introduced by inadequate or biased sampling or other research design flaws. The null hypothesis is the assumption that any connection between the independent or predictor variables and dependent variables is due to chance. Inferential statistics are used to test the null hypothesis. Rejecting the null hypothesis leads to acceptance of the alternative or working hypothesis, that the independent variable does make a difference. The usual probability levels used for rejecting the null hypothesis are the .05 or .01 levels of statistical significance.

A t test or ANOVA is used for testing the significance of differences between means. If you are looking at differences between groups where the measures are categorical, such as observed frequencies or counts, use Chi square.

Correlation answers a different type of question. It indicates the *degree* of a relationship between two sets of scores. A correlation may be either positive or negative.

Box 18-8 lists all of the statistical notations and definitions used in these chapters. Box 18-9 can be used as a guide in selecting an appropriate statistical test for your data.

BOX 18-8. Common Statistical Notations

X	An individual score
Y	An individual score (used for a second set of scores).
N	Number of scores.
Σ	Sum of the term that follows.
ΣX	Sum of all the individual scores.
ΣX^2	Sum of squares. Each score is individually squared and then all the squared scores are added together.
$(\Sigma X)^2$	Sum squared. First, the individual scores are added together; then, this total is squared.
$\overline{X}$	Arithmetic average (mean) of the X scores.
$\overline{Y}$	Arithmetic average (mean) of the Y scores.
SD	Standard deviation, a measure of variability or dispersion, often noted as σ.
t	Test for the difference between two sets of scores with continuous distributions.
ANOVA	Analysis of Variance
F	Ratio of two sample variances
SS	Sum of squares
MS	Mean squares
χ^2	Test for determining the difference between two sets of categorical scores.
p	Probability that the results of a statistical test are due to chance.
$>$	Greater than.
$<$	Less than.
df	degrees of freedom.
r	Pearson product-moment coefficient; a measure of correlation for continuous data.
rho	Spearman rank-order coefficient; a measure of correlation for ranked data.

Does the research question or hypothesis make a comparison among groups?

NO → Does the research question or hypothesis propose a relationship between two sets of scores or counts?

- NO → You have nothing to analyze statistically. Recheck earlier steps.
- YES → Are the measures on both sets of scores continuous?
  - NO → Can the results be ranked from highest to lowest (assigning a rank of 1 to the highest)?
    - NO → Is one variable categorical and the other continuous?
      - NO → Use the Tetrachoric coefficient (r_t) (consult a statistics book)
      - YES → Use the Biserial Coefficient (r_b) (consult a statistics book)
    - YES → Use the Spearman rank-order coefficient (*rho*) (Box 18-7)
  - YES → Are both measures assumed to be normally distributed in the population?
    - NO → Rank the results assigning 1 to the highest → Use the Spearman rank-order coefficient (*rho*) (Box 18-7)
    - YES → Arrange results in pairs and use the Pearson product-moment correlation (r) (Box 18-6)

YES → Are the resulting data (measures of the dependent variable) continuous?

- NO → Present data in a contingency table and use Chi-square (x^2) (Box 18-4)
- YES → Is the dependent variable measure assumed to be normally distributed in the population?
  - NO → Calculate the Grand Median (for all scores) and count the number of cases falling above and below it for each level of the independent variable → Present data in a contingency table and use Chi-square (x^2) (Box 18-4)
  - YES → Arrange data in frequency distributions by group, based on levels of the independent variable. → Are there more than two groups?
    - NO → Are the two groups of scores independent
      - NO → Use the t-test for matched groups (Box 18-2)
      - YES → Use the t-test for independent groups (Box 18-1)
    - YES → Use an Analysis of Variance (Box 18-3)

Reference

Anthony K. (1977). On cigarette smoking—Where do you stand? *GASP Educational Foundation News, 11,* p. 3.

Further Reading

Bruning, J. L., & Kintz, B. L. (1987). *Computational handbook of statistics* (3rd ed.). Glenview, IL: Scott-Foresman.

Fitz-Gibbon, C. T., & Morris, L. L. (1987). *How to analyze data.* Beverly Hills, CA: Sage.

Mohr, L. B. (1990). *Understanding significance testing.* Beverly Hills, CA: Sage.

Runyon, R. P., & Haber, A. (1988). *Fundamentals of behavioral statistics* (6th ed.). Reading, MA: Addison-Wesley.

Siegel, S., & Castellan, J. J. (1988). *Nonparametric statistics for the behavioral sciences* (2nd ed.). New York: McGraw-Hill.

Sample Problems

1. Describe in words the meaning of each of the following statistical expressions.

$t = 2.53$
$N = 20$
$\sigma = 1.9$
$r = .41\ (p < .01)$
$r = .30\ (p < .05)$
$r = .15\ (p > .10)$
$t = 2.53,\ df = 18,\ p < .05$

2. A college instructor wants to see if there is a difference in the final grade between students who participate frequently in class and those who never say a word. In a class of 26 students, the instructor identified 6 participants and 7 nonparticipants. Below is a list of the final grades of these students (A = 4, F = 0). Apply a *t* test for independent measures to see if there is a significant difference between the final grades of high and low participants.

Final Grade (A = 4, F = 0)	
Frequent participators $n = 6$	*Non-participators* $n = 7$
4	4
4	3
3	3
3	2
3	2
1	1
	1

3. Use a t test for *dependent measures* to see if subjects in an attitude change experiment became more favorable to space exploration after seeing a NASA film.

	Attitude scale score	
	Before film	*After film*
Eva B.	62	64
Matt B.	71	75
Gretchen H.	68	67
Cindy H.	51	52
Roger M.	58	61
Margy S.	84	85

4. Is there a correlation between age and speed on a running test? First compute a Spearman rank-order coefficient for the following data using the formula in Box 18-7 and then compute a Pearson product-moment coefficient using the formula in Box 18-6. After you have done this, check the appropriate significance tables in Appendix A to find out whether these coefficients are statistically significant.

Age (years)	*Running speed (in seconds)*
12	20
13	18
13	19
14	14
15	17
16	16
16	13
16	15

Answers to Problems

2. $t = 1.16$, $df = 11$, not significant.
3. $t = 2.33$, $df = 5$, not significant.
4. Spearman $rho = -.744$, $df = 7$, significant at the .05 level (but not at .01). Pearson $r = -.785$, $df = 6$, significant at the .05 level (but not at .01).

19 Calculators and Computers

Calculators

Many makes and models of hand calculators are available. Some are as expensive as a microcomputer and can do almost as much. Others are inexpensive models that can perform simple arithmetic operations but not much else. In between are many low- and moderate-priced calculators that will perform most of the major operations needed for statistical tests. The size of a keyboard doesn't necessarily reflect its potential utility. For most behavioral problems, there will be no need to transform gallons into liters or compute the reciprocals of cosines. Logarithms can more efficiently be looked up in statistical textbooks, on those rare occasions when they are needed in behavioral research, than retrieved on a calculator. We have not found programmable calculators to be especially useful to beginning researchers. At the point when students are ready to write programs, they might be better served by a computer that will open many more avenues for analysis than are available with a hand calculator.

Some features increase the usefulness of a hand calculator. These features are identified by symbols or terms found on particular keys or just above. When a symbol is written above rather than on a key, a separate function button (F) must be pressed first. Symbols vary among different makes or models even when the operations they perform are similar.

Box 19-1 lists useful keys that reflect our experiences with the computational needs of beginning researchers. For those whose work becomes highly specialized, certain keys such as mean ($\bar{x}$) or percentage (%) may increase efficiency, even though these operations can be performed using other keys

BOX 19-1. Helpful Features in Calculators

STO or M	Memory storage. Used for placing items into memory to make them available in subsequent calculation. Several memories are better than one. Essential.
RCL	Memory recall—recalls items placed into memory. Essential.
x^2	Squares each number. Essential.
$\sqrt{x}$	Square root key. Essential.
SS, Σx^2, or $\Sigma+$	Sum of squares. Key used for squaring numbers individually and then summing them. Very useful.
σ, s, or S.DEV	Standard deviation. Very useful.
r or corr	Correlation coefficient or Pearson *r*. Very useful.
t or t_{ind}	*t* ratio for independent scores. Very useful.
t_{dep}	*t* ratio for dependent (correlated) scores. Occasionally useful.
χ^2	Chi-square. Occasionally useful.
F	*F* ratio—used in analysis of variance. Occasionally useful.
Var or s^2	Variance or square of the standard deviation. Occasionally useful.

with little extra effort. The keyboard symbols may vary slightly according to make or model.

Computers

There are three general uses for computers in behavioral research. These functions can be performed without computers, but computers make them easier, faster, and more reliable. The first use, and the one we will emphasize here, is data analysis. Raw data from a research study are entered into a computer and a program is used to compute descriptive and inferential statistics. The second use is word processing. In this case the computer becomes a more efficient typewriter. The third purpose is for running laboratory experiments. A computer can schedule and present stimuli, record, and classify responses. These are referred to as "real-time" computer systems. Setting up a real-time computer system requires more knowledge of programming than we can present here. Good guides are readily available (see the Further Reading section at the end of the chapter).

Finding a Computer

Nearly all universities and colleges, and many other institutions and organizations, such as school districts and corporations, have access to a *main-*

frame. These are large computers (made by IBM, Burroughs, Siemans, DEC) that can process extensive quantities of data in a very short period of time. Terminals connected to the mainframe are located at various offices where data can be entered. You may never see the computer itself, which is probably located in a temperature-controlled basement.

Instead of a mainframe, you are likely to have access to *microcomputers,* also known as personal computers (PCs). These include standard desktop versions and small portable laptop models.

Hardware refers to the physical parts, the actual machine. *Software* consists of programs or packages that direct the computer to perform specific tasks. Software tells the machine what to do. A personal computer with a hard disk, a type of hardware, can accommodate a considerable amount of data and use sophisticated software for statistical analyses. If your computer lacks a hard disk, some of the more complex packages will not run, but there are statistical packages that will run on diskettes—small plastic records that hold software and data.

If you are considering the purchase of a personal computer, read some of the inexpensive paperback books that offer useful hints before you spend your money. You should also talk with friends about their computers, and beware of fancy packaging. The best-looking computer may not be the one you need. You will need to decide on a printer (the typewriter aspect)—letter quality, dot matrix, and laser. A letter-quality printer gives text that is more readable and pleasing to the eye than does a dot matrix. However, dot-matrix printers tend to be less expensive and are able to print graphics. In addition, the print styles (the shape and size of the letters) are practically infinite with dot matrix. If you intend to use a letter-quality printer because you will be working with written text, it will be of no value to buy a computer for which there are marvelous graphics programs—you won't be able to print any of them. Laser printers are the most versatile, combining the desirable features of letter-quality and dot matrix printers, but they are much more expensive. Think through what you want to do with your printer and computer. Find out what programs are available for your purposes. Then select the computer that will run the desired programs.

A major consideration in purchasing a computer is the availability of support—that is, someone available to help you learn how to use it. Try to purchase your computer close to home so your daily telephone calls for assistance will not result in an astronomical bill. No matter how "user friendly" your computer, or how well-written the user's manual, you will need advice and occasional encouragement while getting used to your machine. A local users group can be very helpful. This is an informal organization of people who own your type of computer. When you are considering purchase, ask the salesperson for names of people involved in the local user groups and don't hesitate to contact them. Members are generally very willing to share experiences and offer advice.

Finding a Program for Data Analysis

For a Mainframe Computer

Organizations that invest in mainframe computers are likely to have consultants available to assist potential users. Take full advantage of resource people and materials, such as introductory training packages or published guides, in educating yourself about available programs. There are special programs available for statistical analysis; therefore, you will not need to write your own. Some commonly used programs for analyzing behavioral data are SPSS-X (Statistical Package for the Social Sciences), Minitab, BMDP, and SAS (Statistical Analysis System). Each of these has a published manual describing what the program does and how to use it. They may include a tutorial in the program itself.

For Microcomputers

In selecting a program for a PC look for the key term, *statistical analysis,* in the program title. Data base management and spreadsheet programs might help you organize information, but will be of limited value in statistical analysis. Spreadsheets have been designed for business purposes. Instead, look for statistical packages available for your brand of computer. There are many good programs. The more sophisticated software packages are SPSS/PC+, SYSTAT, BMDP, and SAS for MS-DOS systems (IBM-type PCs). JMP-IN and SYSTAT are available for Macintosh computers. There are also many less elaborate packages that offer statistical capabilities to meet the needs of a beginning researcher. In their descriptions, look for the statistics listed in Chapters 17 and 18.

Each program is slightly different. You will find that once you have learned to use one software package, you will quickly be able to learn another. Most of the software comes with a tutorial—that is, a learning module to familiarize the user with the program. While it might seem tedious, it is extremely worthwhile to go through it at the beginning. You will be saved considerable time and aggravation later by getting an overview of the program's uses, functions, and commands. At the same time, learn the precise meaning of all the terms used. Some of these will be ones that you met in Chapter 6 (Experimentation), in particular, *variable* and *value.*

Analyzing Your Data

Statistical work on the computer has two aspects: data entry and data analysis. The entry and the checking required are the more time-consuming. All the numerical information you wish to use must be correctly entered into the computer and properly saved for subsequent use. Data analysis refers to the statistical summary and tests. These may include both descriptive statistics and inferential statistics (see Chapter 17 and 18).

After you have read about the program and gone through the tutorial, run a test program before you enter your actual data. Make up a list of five or six single-digit numbers, and calculate the sum, mean, and standard deviation (if you have a calculator). Then enter the numbers into the computer, request the statistics just described, and see if you get the correct results. In this way you will have more confidence in the results from your larger data set. Also, it is extremely frustrating to enter 65 multidigit numbers and then discover that you have done it wrong and must redo the entire set. Here is a set to use:

Case No.	*Score*
1	4
2	7
3	—
4	3
5	5

N = 4 (1 missing case)
Mean = 4.75
Standard Deviation = 1.71

Always remember that the computer is a very single-minded concrete thinker that works methodically, step by step. It just does it very fast. If you work step by step, and not so fast, you will be successful in using the machine. All variables are of equal importance to the computer. It makes distinctions only on the basis of your commands. A trial run on either the mainframe or personal computer will save you great amounts of time and effort in avoiding problems that will be more difficult to trace in your actual data.

Data Entry

Codebook

First, make a *codebook*. The codebook, which needn't be a book, is a listing of all your *variable* and *value labels* (levels of the variable). The first variable is likely to be the subject or observation number. You must make up a name for each variable, usually limited to eight letters (five in some programs) without spaces or symbols. This is the *variable label.* For subject or observation number you might use SUBNO, NUMBER, or ID. Even if the program will assign a case number, you should assign your own because if you reorder your subjects or cases, the computer will simply renumber them in consecutive order, thereby losing the original identity number.

The *value* is the number used to describe levels of the variable. The variable GENDER will have two values, male and female. The values of the variable will be the numbers you assign to represent male and female. If male = 1 and female = 2, the value label for 1 on the variable GENDER is male. It is often difficult to recall whether you used 1 for male and 2 for female, or vice versa. An orderly codebook listing variable and value labels, preferably with a copy in a safe place, can prevent frustration when you return to the data at a later date. A codebook is absolutely essential if another person will work on the data. On many projects, data analysis will continue after an employee leaves. The next person will rely entirely on the codebook to make sense of the numbers.

Missing Data

The next task is to learn how missing data are handled. Chances are that in some cases, the value for the variable is unknown. Do not use zero. There are two reasons for this. One is that it is quite possible some of the variables may have a legitimate value of zero. You might have a question about attitudes toward abortion where Agree = 1, Disagree = 0. Here a value of 0 does not represent missing data. You must be consistent in your designation of missing values. Zero cannot be a missing value for one variable and a meaningful value for another. The second reason pertains to the data analysis. When calculating the mean, the program is likely to read the 0 as a score, rather than a missing value, thereby giving you an incorrect result. For example, perhaps your variable is the number of times a person has checked a book out of the library in the past month. If the computer does not specifically know that you are using 0 as a missing value, it will assume that in that case the person did not check out any books, and will give you an incorrect mean as shown in the third column below.

SUBNO	(correct) NUMBOOKS	(incorrect) NUMBOOKS
1	3	3
2	8	8
3	1	1
4 (missing)	—	0 (incorrectly read)
5	2	2
Mean =	14/4 = 3.5 (1 missing)	14/5 = 2.8

SUBNO = subject number
NUMBOOKS = books checked out

Now that you have been convinced not to use 0 for missing data, what should you do? Most programs have a specific value or symbol for missing data. For example, SYSTAT uses a period (.). When the program encounters the dot (.) it knows not to include the case in the analysis, and, instead, counts it as missing. Other programs will allow you to choose a value to represent missing data. In this case using -99 or -999 is a safe procedure. If it is inadvertently used in the statistical analysis, you will get such a peculiar answer as to know that something is wrong. Using the library book example:

SUBNO	NUMBOOKS
1	3
2	8
3	1
4 (missing)	−99
5	2
Mean =	−85/5 = −17 (weird answer waves a red flag)

SUBNO = subject number
NUMBOOKS = books checked out

Backup Copies

We cannot stress enough the importance of making a backup copy of your data entries on a diskette that you keep yourself. Unimaginable things can happen to data you have entered into computer memory. Expect it to be lost, either from your own or some external cause. Extremely hot weather can lead to extensive use of air conditioning, which in turn can lead to power fluctuations in an entire city that may affect your individual computer. If you share computer use, there is little control over the activities of others. Panic and paranoia are easily avoided by having a personal copy of your data set.

Data List

After the raw data have been entered, print out your entire data file. This is called a *data list.* Use this to check and double check your entries. Keep it. If all else fails (i.e., your file disappears and your backup diskette melts) at least you will have a typed copy. It is tempting to rush ahead to the final analysis, but save paper and first make sure all your entries are correct. It is best to read your raw data aloud to another person who checks the entries on the data list. After you make changes, be sure to recopy the corrected file onto your backup diskette.

Data Analysis

Once your data entries are checked and corrected, the analysis is amazingly fast. Following the instructions of your particular program, you simply request the statistics you want, and answer the computer's queries about which variables you wish to use. Here the discussion in Chapter 6 on Experimentation will be helpful. You will need to specify which variable represents the outcome, or dependent variable, and which variable or variables comprise the independent or predictive ones. Sometimes these will be referred to as the grouping variable or groups, as when comparing one group with another on some outcome.

Look at the results and see if they make sense. If the scores range from 40 to 100, the mean cannot be 22.57. Check to see that the number of cases (N) corresponds with your knowledge of the number in your sample. If you have been paying attention to your data during the study, there should be no big surprises in the data analysis.

Label your output. Write notes on the printout indicating the date and type of analysis made. Some computer programs allow you to insert value labels that then appear on the output; however, others do not, or you may not wish to take the trouble of using that aspect of the program. In such cases, you must add the information by hand. Otherwise, you may look at the printout a few days later and not have any idea which analysis it represents. Once you gain an understanding of the process, you are likely to start making a number of subset analyses (i.e., comparing female Protestants under the age of 30 with male Catholics over 55). Keeping a record of your computer runs with the dates is useful.

BOX 19-2. Steps in Data Entry and Data Analysis

Data Entry

1. Read over the material accompanying the statistical package you will be using.
2. Go through the tutorial.
3. Run a test data set, including an entry with missing data.
4. Resolve any problems by either going to the written material with the package, the tutorial, or asking a knowledgeable person.
5. Make a codebook.
6. Enter your raw data.
7. Make a backup copy of your data on a separate diskette.
8. Print out a data list, adding label and date.
9. Thoroughly check the data list for errors, circling corrections in red.
10. Correct errors and make a new backup copy.

Data Analysis

1. Request the desired statistical analyses.
2. Label the output, adding date, time, variable and value labels, as needed.

Data Presentation

Behavioral science journals publish most data in tables. With a good software package, it should be possible for you to print out your tables in journal style. Most statistical software packages also have a graphics capability, allowing you to present some of the data in visual form. This may be useful in showing relationships that have a particularly interesting and clear form. With a few adjustments it is possible to make tables, charts, or both from the same data set. The program manual will explain the construction of graphs and tables. Don't overwhelm yourself or anyone else who reads the paper with duplicate tables plus graphs for every single comparison.

Avoid Blind Analysis

The computer has greatly extended the range of statistical techniques available to researchers. It can perform operations in a matter of seconds or minutes that would take hours or days to analyze with a hand calculator. The positive effect of this has been to broaden the range of methods of data analysis available, but the new freedom of choice brings the risk of asking the computer to perform analyses that the researcher does not understand. It is unwise to take a complicated printout of data to a statistical consultant in the hope that he or she can make sense of it. The consultant may be familiar with the statistical test but not the nature or limitations of the data being analyzed.

As a general rule, a researcher should not travel too far beyond the bounds of personal understanding in doing a statistical analysis. The problem was less serious with hand calculation where one could gain familiarity with a technique in the course of doing the computations. Today, a computer does everything so quickly that there is little time for understanding to develop. You should read about a statistical method before asking the computer to perform it. This doesn't mean that you have to understand the derivation of all the formulas in order to apply a statistical test. What is needed is a general understanding of the nature of the test and the assumptions behind it, what it can and cannot do, and the meaning of key concepts and terms. This understanding can be gained from reading some of the statistical textbooks listed in Further Readings at the conclusion of Chapter 18.

Finding Help

Many college campuses offer courses for students and staff on the use of computers and on managing data files. Extension courses are also common. While you may not find that all the information is pertinent to your interests, such classes often provide valuable time and money-saving tips.

Chances are that your institution or organization will have some trained computer specialists on the staff. They are there to help you. There are also a number of paperback books available in the computer section of bookstores. As these may be overwhelming in number, talking to friends and colleagues may be an easier way to get assistance.

Summary

For data analysis, select a calculator that has some memory capacity and keys for calculating square roots and for squaring numbers. Having the capacity to accumulate the sums of the squares is very useful.

Two general types of computers are mainframe and microcomputers. If your institution has a mainframe computer, it is likely that consultants are available to help you. Commonly used data analysis programs are SPSS-X, Minitab, BMDP, and SAS.

Desktop microcomputers are useful for analyzing data from small- to moderate-size studies. If you decide to purchase a computer, consult other users and read books to familiarize yourself with what is available. Carefully think through your research needs. Find out what programs exist to meet those needs, and then buy the computer that will run them. Letter-quality printers are preferable for written work, but dot-matrix printers may be acceptable, are generally less expensive, and have graphic capabilities.

Run a small dummy set of data in order to understand exactly how the program works. A trial run will assist you in setting up your data collection and coding to facilitate computer analysis. Be sure to have a clear codebook listing of all variables and value labels. Make backup copies of your data entries and update them when you make additions or corrections. Label and date all printed output.

Avoid blind analysis. Be knowledgeable about the specific statistical techniques you are using.

Further Reading

Bear, John (1983). *Computer wimp: 166 things I wish I had known before I bought my first computer.* Berkeley, CA: Ten Speed Press.

Bird, R. J. (1981). *The computer in experimental psychology.* London: Academic Press.

Cozby, P. C. (1984). *Using computers in the behavioral sciences.* Palo Alto, CA: Mayfield.

Deni, R. (1986). *Programming microcomputers for psychology experiments.* Belmont, CA: Wadsworth.

Dixon, W. J.; Brown, M. B., et al. (1985). *BMDP statistical software.* Berkeley, CA: University of California Press.

Kahn, E., & Seiter, C. (1985). *The skeptical consumer's guide to used computers.* Berkeley, CA: Ten Speed Press.

Norusis, M. J. (1988). *The SPSS guide to data analysis for SPSS/PC+.* Chicago: SPSS, Inc.

SAS Introductory guide for personal computers. Cary, NC: SAS Institute.

Wilkinson, L. (1987). *SYSTAT: The system for statistics.* Evanston, IL: SYSTAT, Inc.

20 Writing and Reviewing a Research Report

Set aside a special time for writing and editing. Don't expect to complete the analysis of your data and hand in the final report a few days later. Several drafts will probably be required. It is important to get feedback from other people. Early drafts should be labeled as preliminary and dated.

There are two general classes of reports. One is an article for a scientific journal and the other is a technical report written for a specific client or audience. The journal format is easier to use for hypothesis-testing studies. Opinion or attitude surveys and evaluation studies are more easily written in technical report form. The rules for the form of a scientific article are quite precise. Technical reports can be much more varied in style and order of presentation. The decision regarding the form of data presentation will depend on the type of dissemination you want. If you desire to write articles for other researchers, you will have to learn journal style.

Article for a Scientific Journal

Many scientific periodicals require that contributors follow a specific style. This is usually mentioned in the inside front cover under the heading "Instructions to contributors." When you write for a specific journal, be sure to check these instructions before beginning the report. The American Psychological Association (APA) publishes a style manual for its journals. This manual can be ordered directly from the American Psychological Association, 1200 17th Street N.W., Washington, D.C. 20036. APA style requires that submitted articles must be double-spaced, typed on one side of the page only, and follow certain procedures such as placing a "running head" (abbreviated title) on the top right of each page. This standardized format is intended to assist the editor and reviewers since all of the sections will be in predictable places. When the paper is finally published, it will be printed single-spaced on both sides of the page, and often with two columns to economize on valuable journal space. It is the submission format that is described in this chapter. Appendix B contains an experiment written in APA style.

Because of the importance of experiments in the history of psychology, APA format works best for that particular method. Since students in research methods courses often do nonexperimental projects involving questionnaires, interviews, and observations, it is important to know how to adapt these studies to APA style. Our experience has been that most problems with writing reports of nonexperimental studies occur in the Methods section since there is no equipment or subjects in the narrow sense, but participants, respondents, or simply people who happened to be in a place during the researcher's visit. An example of the Method section of an observational study written in APA style is shown in Appendix C.

The layout of a manuscript prepared in APA style includes the following sections: Title page, Abstract, Introduction, Method, Results, Discussion, and References. Box 20-1 contains a checklist of these items.

BOX 20-1. Checklist for Writing a Scientific Article

General format

- Typed doubled-spaced with 1½ inch margins.
- Type title page, Abstract, References, and Tables, on separate pages.

Title page

- Title summarizes the main idea of paper.
- Name and institutional affiliation of author(s).
- Running head—abbreviated title no more than 50 characters.
- Acknowledgments of assistance and research support.

Abstract

- Specify subject characteristics (number, age, sex, etc.).
- One or two sentences describing the problem, the method, and findings.

Introduction

- Opening paragraph tells what the study is about, and introduces the problem.
- Subsequent paragraphs describe background material and previous studies.
- Specify hypotheses or questions underlying the research, and present rationale for each.

Method
Subjects
Number and method of selection, characteristics (age, sex, etc.), procedures for handling refusals or nonreturns.
Apparatus (or *materials*)
Description of observation sheets, tests, scales, or measuring devices.
Methods of development if novel instruments were employed.
Coding and scoring procedures.
Procedure
Experimenters/observers/interviewers—relevant characteristics, training.
Reliability of procedures and instruments.
Setting—location and relevant environmental characteristics (lighting, sound, temperature, etc.)
Instructions to subjects/respondents.
Duration and number of sessions.
Control features such as randomization or counterbalancing.

Results
Begin with most relevant findings.
Describe results in quantitative way (numerical) and provide necessary descriptive statistics (means, standard deviations, contingency tables, etc.) along with the inferential statistics.
Make a summary statement about these results.
Properly label all tables and figures.
Tell the reader what to look for in tables or figures.
Describe qualitative findings that are pertinent and important.

Discussion
State whether hypotheses were supported or rejected, or address questions raised in the introduction.
Discuss similarities and differences between current findings and earlier ones.
Describe the implications of the findings.
Describe ways of improving the research, unanswered questions, new questions raised, and directions for future research.

References
List all citations that appear in the text.

Title Page

This contains the title of the paper, authors' names, their institutional affiliation (university, private firm, government agency, etc.), and the running head, which is an abbreviated title printed at the top of each page on the typed manuscript. The title should be fairly short, perhaps 12–15 words, summarizing the main topic or focus of the paper. Avoid unnecessary words in the title, such as "An Experimental Study of . . ." or "A Research Investigation into . . . ," which take up space and don't add any content. Examples of suitable short titles would be "Achievement Motivation in Children" or "The Effects of Amphetamine on Reaction Time." The running head should be a maximum of 50 typewritten characters counting letters, punctuation, and spaces. An example of a title page for a submitted article, plus all the other sections, is found in Appendix B.

Abstract

Because this is all of the report that most people are likely to read, it is very important that the abstract be clear, concise, and accurate. Its length will depend on the publication or distribution. The abstract of an article for a scientific journal should not exceed 150 words. Someone reading an abstract should be able to see at a glance the area covered by the study and the main findings. The reader who wants more information should consult the body of the report. In practice, the abstract is written after the rest of the report is completed.

Introduction

The introduction begins with a statement of the problem, its theoretical and practical significance, and its place in a larger body of knowledge. The introduction includes a review of related studies and a list of specific hypotheses or predictions.

Method

This section describes specifically what was done. The interested reader should be able to repeat your study from the account provided here. Included should be a description of the apparatus, tests, or observation sheets, the subjects and how they were selected, the interviewers and their training, a description of the physical setting in which the data were collected, instructions to the subjects, and mention of any special problems that arose. Describe any pilot tests or reliability checks made on the procedure or equipment.

Results

This is the place for factual information about what was found. Opinions and interpretations are reserved for the discussion. Begin the section with the findings most relevant to the hypothesis or problem. Then present secondary or related findings. The order of presenting results should follow the order of the hypotheses in the introduction. Don't skip from hypothesis 1 to 4 and then back to 2. Let the reader see how each hypothesis fared before going on to the next.

Do not overwhelm the reader with raw data, particularly if you are also presenting summary statistics such as percentages, averages, and probabilities. Generally, you will be able to present the necessary summary data in the text or in tables.

There is a widespread belief that "some people read numbers; others narratives. No one likes both" (Toch and Adams, 1988, Pg. ix). This statement provides little guidance in deciding how many numbers are to be included. Sometimes the issue can be resolved by consulting the journal in which you want to publish your results. If the journal tends to publish articles containing many detailed statistical tables, you will probably want to emphasize the quantitative aspects of the data. The reverse would be true if the journal prefers qualitative description. In cases where there is a choice as to whether to include numbers or words, we recommend some degree of redundancy (repetition). However, minor information from the tables, such as the means and standard deviations for unimportant comparisons, do not require elaboration in the text.

Tables and Figures

If they are needed, most tables and figures will be presented in the Results section. *Tables* refer to tabular information that can be typeset, such as lists of terms or columns of numbers or percentages. *Figures* are charts, graphs, drawings, photographs, or any other type of pictorial information. Figures are more time-consuming and expensive to prepare and reproduce than either text or tables. Each table and figure must have a numbered descriptive title, for example, "Figure 1. Creativity Scores by Respondent Age." The sample report in Appendix B contains an example of a table. Figures (and tables) do not speak for themselves. Many readers automatically skip them. Therefore, it is essential to provide a written description as well.

If you are considering tables and figures, use the following questions adapted from the Publication Manual of the American Psychological Association (1983).

Is the table necessary?
Does every column have a column heading?
Is it simple, clean, and free of extraneous detail?
Is it easy to understand—is its purpose readily apparent?
Are the titles brief but explanatory?
Is the table or figure referred to in the text?
Does the text include an indication of placement for each table and figure?

Figures submitted for publication need to be photographed on a glossy print, identified on the back lightly in pencil, and accompanied by the numbered caption typed on a separate page.

Discussion

This is your opportunity to interpret the findings, discuss their significance, and suggest directions for future research. There is no need to repeat everything that was said in the results. Instead, describe how your findings fit existing theories and other research in the field. What general conclusions can you draw? Based on what you have learned, what issues should be investigated next? What are some of the flaws and limitations of the study? If you were to repeat the study, what would you change?

References

This is a list of materials cited in the body of the report. It does *not* include relevant work not specifically mentioned. A long list of references is no mark of scholarship if they have come from secondary sources (accounts about other people's work) and you have not read the original articles. Reference lists should be triple checked. It is frustrating to try to locate an article that is supposed to be in Volume 12 and find that it is in Volume 21, or that a book author's name is Hemminski rather than Chemminski. The specific form for the references will be determined by the periodical or audience for whom you are writing. The American Psychological Association (APA) style manual requires the following form for citing articles.

Example of an article citation:

Smith, R. S., & Jones, D. A. (1991). Children's response to colored glasses. *Journal of Child Development, 30,* 14–26.

Example of a book citation:

Brown, O. (1989). *Organizational behavior.* New York: Smith Publications.

The American Sociological Association (ASA) requires a different reference style for its major research journal, the *American Sociological Review.*

Example of ASA style for an article citation:

Robinson, R. A. and Philip Agisism
1951 "Making mail surveys more reliable," Journal of Marketing 15:415–24.

For a book citation:

Ornstein, M. D.
1976 Entry into the American Labor Force. New York: Academic Press.

Technical Reports

Often a study is written for a specific client, such as an organization, firm, or public agency that sponsored the study. It may depart from the scientific journal format. Some clients want everything included, others want only brief accounts of the major findings. Don't hesitate to check the form, style, and length used by others who have written similar reports. Consider this part of the literature review. You won't need to follow the style used by others exactly, but it is important to know how they have presented their work. Two widely used style manuals are *The Chicago Manual of Style* (University of Chicago, 1982) and *The U.S. Government Printing Office Style Manual.* An example of a technical report can be found in Appendix D.

The length of the report depends largely on the wishes of the sponsor and the projected audience. The decision involves a compromise between presenting sufficient information so that the findings and basis for the recommendations are clear but not given in so much detail that people will be discouraged from reading the document. A frequently used way of getting the point across, while providing sufficient backup information, is to begin with an *executive summary.* An executive summary is as brief as possible, but covers all the necessary information—the problem, method, results, and conclusions.

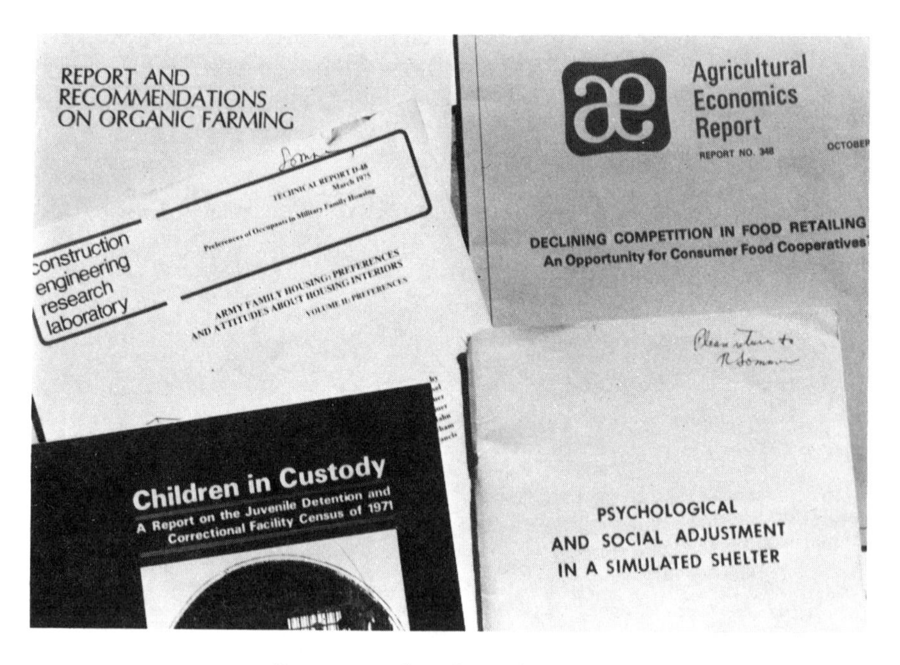

Some examples of technical reports.

Photographs and drawings can be used to liven up a technical report. A study of housing needs of the elderly might include pictures of senior citizens and various housing options. Large organizations such as government agencies or corporations that commission numerous printed documents may require a specific format and style for all technical reports. The report may use a standard cover displaying the corporate logo and statement of endorsement by an executive officer. Advance approval may be necessary before a report is officially released. Prior to writing a technical report for a sponsor, check in advance about approval policies.

Appendix

A technical report is likely to conclude with an appendix section. This is the place for all supplementary materials not included in the main body of the report. Appendices are arranged in sequence with letters (e.g., Appendix A, Appendix B, etc.). The appendix contains material that is not essential for understanding the report. It will be used most often by readers with a special interest in the topic who want detailed information. Typical items considered for inclusion in the appendix are:

1. Score sheets, questionnaires, or observation sheets used in the study. An exception would be instruments that are commonly used and widely available. These can be omitted.
2. Detailed tables and charts primarily for reference purposes. Tables and figures necessary for the reader's understanding should be placed in the results section.
3. Lists of technical terms. For a general audience, technical terms should also be defined as they occur in the text.
4. Other relevant documents not essential to the reader's comprehension of the report. These might include the sign-up sheet for recruiting subjects, interviewer instructions, instructions for coding interview data, the follow-up letter to respondents who did not reply, etc.

Writing a Critique

The preceding section of this chapter concerned the written form of a research paper. Another writing task in science involves the preparation of a *critique* or critical review of a single paper or series of papers on a specific topic. Despite its semantic resemblance to negative criticism, the tone of a critique may be positive, neutral, or unfavorable. The critique is central to the *peer review* of scientific work. This is the process by which submitted articles are evaluated by those knowledgeable in the field (i.e., scientific peers). Most journals in the behavioral sciences require that an article pass through peer review before it is published. This can be a long and tedious process in which numerous requests for revisions are made before the article is accepted. Some editors remove all identifying information on the authors from a paper before it is sent out for review. This is called a *blind review* system. Box 20-2 shows the evaluation forms used by two behavioral science

---

BOX 20-2. Examples of two evaluation forms for journal reviews.

Title of Paper and Author:
Reviewer Number:
Please fill out the following, in addition to your typed comments.

1. Style. Is the writing clear and the statements of theory or results unambiguous?
yes______ no______
If no, please give guidance for rewriting.
2. Literature review. Does the paper mention the necessary references?
yes______ no______
If no, please list the omitted ones in your comments.
3. Methods. Have the authors used the best methods available?
yes______ no______
If no, please comment.
4. Interpretation. Have the authors given the appropriate interpretation to their data? yes______ no______
If no, please comment.
5. Significance. Is this paper a significant contribution to its field?
yes______ no______
If no, state reasons.
6. Recommendations:
Should be published, top priority
Should be published, no revision necessary
Should be published with revision
Should not be published

Manuscript Review Form

Manuscript Number: Reviewer Number:
Manuscript Title:
Please review this manuscript within the next 30 days, providing us with your objective evaluation of its publishability.

Recommendation:
Publish as is
Publish with noted revisions
Not publishable in present form
Better suited for another journal (specify name)
Not publishable

General Comments for Editor/Author

---

journals. The first journal requests its reviewers to rate the paper along specific dimensions before coming up with a recommendation, while the second one requests a recommendation followed by written comments.

Somtimes a critique of a published article is part of a class assignment. When making your critique, keep in mind that most journals put severe space limitations on authors. A master's or doctoral thesis is likely to run 50–200 pages, but a journal article describing the identical study may be only a fraction as long. This means that relevant details will be omitted from the published version. Recognizing that even key information is missing, there are still criteria you can use to evaluate the quality and adequacy of a scientific paper. These are summarized in Box 20-3 in the form of questions that you can ask as you read through an article. These questions will also be helpful in preparing your own reports.

BOX 20-3. Checklist for reviewing a research report

I. Introduction and background.
   A. Is the problem clearly stated?
      1. What questions are posed by the researchers?
      2. Are the hypotheses or goals of the study clearly stated?
   B. Is the important background literature (earlier work in the field) included?

II. Research methodology
   A. Are the methods appropriate to the goals of the study?
      1. Do they fit the question?
      2. Was the choice of methods dictated by the goals of the study (the problems), or simply a matter of expedience (the course of least resistance)?
   B. Is the sample appropriate, given the nature of the study (i.e., is it representative of the population of interest)?
      1. Were the participants selected in such a way as to avoid bias?
      2. Is it a special group whose characteristics may influence the outcome in particular ways?
      3. Does the sample make sense for the hypotheses being tested?
   C. What outcome is being measured?
      1. How is it measured; for example, a paper-and-pencil test, an open-ended questionnaire, a time measure, counts of some activity, statistics from a government report, etc.?
      2. Are questionnaires, surveys, or tally sheets described in sufficient detail for the reader to judge their adequacy?
      3. Are laboratory instruments adequately described?
      4. In observational studies, is the selection and training of observers described?
      5. Is evidence of reliability presented (i.e., reliability of instruments, interobserver reliability, test–retest information, etc.)?
   D. Does the procedure do what it is supposed to do?
      1. Is the logic of the procedure correct?
      2. Were all sources of bias (subject, experimenter, responder, interviewer, setting, etc.) eliminated or controlled?

III. Description of findings
  A. What answers were obtained to the questions posed?
    1. Are the findings clearly presented with adequate descriptive statistics (e.g., number of cases, mean, median, standard deviation, or range)?
    2. Are the statistical tests (inferential statistics) appropriate to the data?
  B. Are the findings correctly interpreted?
    1. Are the conclusions based on the results?
    2. Is the level of generalization appropriate?
      a. Is it justified by the sample?
      b. Is it justified by the research design?
    3. Are procedural weaknesses noted? Are the limitations of the study discussed?

Need for Empathy

Doing research yourself provides a good perspective for evaluating the work of others on the same topic. This is the basis of peer review, in that experienced researchers are likely to be familiar with previous work on the topic. They also are aware of the difficulties in studying certain issues. Personality researchers know the problems in securing honest responses to personal questions. Experimentalists realize that some subjects won't show up for their scheduled appointments or will incorrectly perform their assigned tasks, leaving blank spaces in the data. Psychophysiologists know how time-consuming and expensive it is to simultaneously record physiological and psychological processes. They understand why a study of drug effects might use only 12 subjects. In most social psychological experiments, this number would be too small for valid generalizations.

A review should enlarge the reader's understanding of the good and bad features of a study. Some critiques are unbalanced in that they provide no guidance to the author as to how the paper could be improved. When you review a paper, ask yourself what new information you are providing that will be of assistance to the author. This is in addition to all the specific comments you've made about the clarity of the abstract, the appropriateness of the method, and so on.

Helpful Hints

Avoid plagiarism by placing quotation marks around any material that is quoted verbatim. If the quotation is extensive, permission to quote is likely to be required by the holder of the copyright. The exact length allowed without specific permission varies. The American Psychological Association allows quotations of up to 500 words from their journals without explicit permission. All direct quotations must include a reference citation with the page number of the quoted material. Using drawings, figures, graphs, or tables from published work is also likely to require permission. Omit quotation marks with material that is paraphrased (expressed in your own words rather than the author's), but be sure to mention the author's name.

Avoid footnotes wherever possible. Footnotes break the continuity of a passage, and many readers will ignore them. Long footnotes that contain detailed explanations are rarely appropriate. The information should be included in the body of the report or omitted.

Double check all the tables for misprints. See that the percentage columns add up to 100. Check that the graphs and figures are titled and inserted in their correct locations. Also, check to see that the vertical and horizontal axes of graphs are labeled correctly. Double check the names of people cited in the literature review and triple check the reference list at the end of the report.

Be sure to show a draft of your report to other people before the final version is typed. Discussing it with people not directly involved with the issues may provide fresh ideas.

The style of journal articles tends to be impersonal and the use of first person pronouns such as "I" and "we" is discouraged. Refer to yourself as "the researcher" or "the investigator." Words such as "data" and "criteria" are plural; the singular forms are "datum" and "criterion," respectively. A common error is incorrect placement of the apostrophe. Note the differences between the simple plural (subjects), the singular possessive (subject's), and the plural possessive (subjects').

In preparing a critique, indicate how the study and write-up could be improved. Be sure to described the good points as well as the deficiencies.

Summary

Journal articles are written in a specific style. In APA style the sections are abstract, introduction, method, results, discussion, and references. Technical reports are more variable in format and often begin with an executive summary. In writing papers it is necessary to avoid plagiarism and provide credit to others whose work is used. Peer review refers to a critique by someone knowledgeable in the field. In preparing a critique it is important to provide positive suggestions for improvement as well as criticism of weak points.

References

American Psychological Association (1983). *Publication manual of the American Psychological Association* (3rd ed.). Washington, DC: Author.

Toch, H. & Adams, K. (1989). *Coping: Maladaptation in prisons.* New Brunswick, NJ: Transaction Books.

U.S. Government Printing Office Staff. (1986). *Style manual* (rev. ed.). Washington, DC: U.S. Government Printing Office.

University of Chicago Press. (1982). *The Chicago manual of style* (13th ed., rev.). Chicago: Author.

Further Reading

American Psychological Association. (1983). *Publication manual of the American Psychological Association* (3rd ed.). Washington, DC: Author.

Day, R. (1988). *How to write and publish a scientific paper* (3rd ed.). Philadelphia: ISI Press

Houp, K. W., & Pearsall, T. E. (1988). *Reporting technical information* (6th ed.). New York: Macmillan.

Jolley, J. M.; Murray, J. D., & Keller, P. A. (1984). *How to write psychology papers: A student's survival guide for psychology and related fields.* Sarasota, FL: Professional Resource Exchange.

Lannon, J. (1988). *Technical writing* (4th ed.). Boston: Little, Brown.

Rosnow, R. L., & Rosnow, M. (1986). *Writing papers in psychology: A student guide.* Belmont, CA: Wadsworth.

Zeisel, H. (1985). *Say it with figures* (rev ed.). New York: Harper & Row.

21 After the Report

Jason Tyburczy, an undergraduate psychology major, studied human–animal interaction in a veterinary hospital waiting room. It was an interesting setting where humans, dogs, cats, birds, and occasional rodents and snakes coexisted in very close quarters. Tyburczy observed pet owners and interviewed them about their level of satisfaction with the visiting area. Having completed the study, he was faced with the question of what to do with the findings. He mailed a copy of the report to the hospital director. No response came back. This lack of spontaneous feedback is fairly typical among recipients of survey results.

Tyburczy next considered various publication outlets. He went to the university library and consulted a reference librarian. He described the study and his ideas about who might be interested in the results. The librarian suggested a number of periodicals. Tyburczy looked at copies to find out which were the most suitable. After some deliberation, the choice narrowed to either a national or a state veterinary journal. The chances of publication seemed better in the state journal so this was the outlet selected.

The report was rewritten in a style appropriate for the periodical and submitted. A few months later, a favorable review was received along with a request for revision. The revised paper was eventually accepted by the journal and appeared in print 7 months later (Sommer & Tyburczy, 1978). A copy of the published article was sent to the hospital director and this time a formal letter of thanks was received. Apparently the published article was more impressive than the typed draft report. A year later, the director asked the researcher for a copy of the findings to be used in a proposed renovation of the waiting area. Several additional requests for reprints and further information about the study were also received.

Further Research

A good study will probably raise more questions than it answers. What begins as a small investigation can easily mushroom into a large research program requiring a major commitment of time and effort. Procedures developed in study may be extended to new classes of problems or people. Or the original sample can be retested using different procedures.

Beginning researchers tend to think in terms of single studies. Experienced researchers think in terms of larger programs in which several related studies are linked together. A research program is economical in time and effort since later studies build on earlier work. A follow-up study is also the opportunity to correct the mistakes of an earlier investigation. There is an old saying among researchers that a first study is *always* flawed. Beginning researchers may not realize this if they read only published articles in which everything seems correct. Many published papers are not first studies but third or fourth attempts to wrestle with a difficult and complex problem. The final version looks methodologically correct because the studies were repeated several times and the manuscript underwent numerous revisions prior to publication.

Journal Review

The review of papers by scientific journals is a time-consuming process. A typical sequence of events is shown in Figure 21-1. For a paper submitted to a behavioral science journal, it is likely that editorial review will take 3–9 months, not including needed revisions, or rejection and resubmission to another journal, which involve extra time. If a paper is accepted, there is

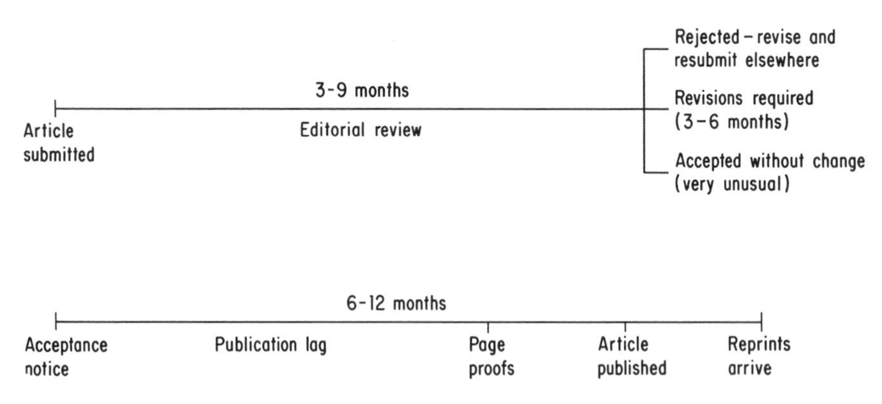

FIG. 21-1. Timetable for publication in a scientific journal.

likely to be a 12-month delay before it is finally published. This may seem a long wait but it is typical of science. Experienced researchers deal with the numerous delays by conducting several studies that overlap in time. While one study is being reviewed, the researcher is working on others which are in various stages of completion. There are also less time-consuming types of dissemination, such as presenting the results at a professional meeting.

Choosing a Journal

Discuss possible publication outlets with those experienced in the specialty area. Also see where related articles have been published. Then visit a university library and look over the most likely prospects. In some fields there is a wide choice among publication outlets. There are more than a dozen English language journals that carry papers on social psychological topics. The same is true of journals in animal behavior. However, in some specialty areas, such as program evaluation or psycholinguistics, the choice among potential outlets is more limited. Check the table of contents and the editoral policy statement of a journal before considering a submission. Journal titles can be misleading, as in the case of *Psychological Bulletin,* which carries lengthy reviews, while the *Psychological Review* publishes mainly theoretical articles, and *Contemporary Psychology* publishes book reviews, only on request from the editors. Often a beginning researcher is wise to choose a less prestigious journal with a lower rejection rate than a top journal that rejects 9 out of 10 papers.

Paralleling the distinction between basic and applied research made in Chapter 1, some journals specialize in basic research while others emphasize applied studies. Because they publish different types of articles, they reach different audiences. Before choosing a journal, check where it stands on the basic-applied continuum. Otherwise you risk having an article rejected by a basic science journal as "too applied" or rejected by an applied journal because it is "too theoretical." You can make this judgment by looking at the articles published. Another strategy is to examine a bibliography on your topic, or your own reference list, and see which journals are represented. If a journal has published several papers on a topic, the editors are likely to be receptive to further studies that extend the earlier work.

Submit a technical article to only one journal at a time. Many journals require a statement to accompany each submission stating that the article is not being submitted anywhere else.

Some journals have submission fees; others charge authors of accepted papers according to article length. For some journals, these charges may be waived for students and others without support funds, and other journals charge no fees. This information is available in the Instructions to Authors section of the journal, which is often found on an inside cover.

Drafts

It is standard practice to circulate preliminary versions of papers to colleagues for their comments. Their input will be invaluable in preparing the final draft for submission. Date the cover page of a preliminary draft and write at the top "Draft copy—for comments only" or "Draft copy—not to be cited." This will indicate to the recipient that the contents are subject to revision before the final draft is submitted.

Dealing with Rejection

The rejection rate among major behavioral science journals is high. For some journals it can reach 80–90 percent of all submitted papers. When a paper has been rejected, it can be revised on the basis of reviewer comments and sent to a second journal for consideration. It is inappropriate to submit an article to more than a single journal at a time. Develop back-up plans for a second or third journal in case the first choice rejects the paper.

Preprints

When an article has been accepted for publication, this can be noted on the cover page of the final version with the notation "In press, Journal of ———." It is legitimate to send copies of an accepted final version (this is called a *preprint*) to interested colleagues who would otherwise wait many months to see it published. Recipients of an "in press" article will be able to cite the article in their own writings. An article formally accepted for publication has virtually the same status and legitimacy as a published article. Note that an article "in press" has been formally accepted for publication, while one "in preparation" is still in the process of review or revision.

Page Proofs

A journal editor may make minor editorial changes on an article accepted for publication. Any changes will be indicated on the page proofs sent to the senior author. This will be the last opportunity to correct typographical or other errors. This is *not* the time to add new material to an article that has already been accepted. Editors discourage this practice and the journal may charge a stiff fee for each line of page proof altered by the author.

Reprints

Most journals do not require an author to purchase reprints. This is an optional matter and depends on the resources available, cost, potential demand, and so on. Some journals supply a small number of free reprints to authors. Other journals charge for all reprints. They may also allow authors to photocopy their own articles and distribute photocopies in lieu of buying reprints. Other journals do not allow this practice. Check the policies of the journal in regard to authors' rights to photocopy their own papers. Ordering reprints is generally done on a separate form that accompanies the page proofs.

Application

The goal of a study may not be further research but application, such as using the information to change policies and practices in an organization. When application is the researcher's goal, a scientific article by itself may be insufficient for changing long-established attitudes and procedures. Consider writing an article for a periodical directed specifically to practitioners. Most occupations, professions, and specialities have periodicals concerned with applied issues. Your reference librarian can help you locate appropriate outlets for applied studies. Such outlets will require a more lively writing style and format than is customary for a scientific journal. However, even in an applied article, busy administrators may not see the potential usefulness of the findings. A common response among administrators to research findings is that they are interesting but not directly relevant. There are always reasons why things should be left unchanged. The researcher may have to convince busy administrators of the value and relevance of the findings through personal discussion.

Application is also an opportunity to learn the limitations and shortcomings of a study. The researcher becomes aware of questions that should have been included on the survey but were omitted, or questions that yielded ambiguous answers. This may demonstrate the need for further studies on the topic to pinpoint problems and develop solutions.

It is difficult to predict what parts of a study will be most useful in applied settings. In several of our surveys for nonprofit organizations, it has been the demographic information that yielded the most interesting information from the client's perspective. Surprisingly, many organizations and agencies do not have an accurate idea of who they serve. As an example, a class project involved interviews with customers at a second-hand store that does an excellent job in rehabilitating the disabled. The management believed that most of the customers had large families. The survey showed the opposite to be true. A possible explanation is that large families have more opportunities than small families for hand-me-downs. This information on family size, thought to be of only incidental interest by the research team, helped the store to define its customer population and improve its outreach program.

Box 21-1 shows the correspondence between the instructor who supervised the class project and the regional chief executive officer (CEO). The first letter illustrates the form of a cover letter for a research report in an applied study, and the replies give detailed feedback on the utility of the findings to the client.

BOX 21-1. Correspondence about a research report

Dear (CEO):

I am pleased to enclose the report on the survey undertaken by the students last quarter. As you will see, the survey shows . . . (description of major findings presented for emphasis).

We hope you find the report informative and helpful. This was a very good learning experience for the students. Sandy, Janee, and Elizabeth felt that they received excellent cooperation from the store and the customers, and appreciate the opportunity to put their research methods training to practical use. We look forward to your comments on the recommendations.

Sincerely,

Professor

Dear (Professor):

I reviewed the study your students made at our store with my staff and we were quite surprised with many of their findings. Without the survey report, we would still be unaware of many factors that are considered to be of prime importance in our operations. First of all, let me inform you that we received many compliments from our customers on the method of conducting the survey. There was not a single negative reaction by any of our customers.

(Detailed reaction to survey results—six paragraphs!)

For the first five months of the year, our sales have been $63,325 over the same period last year. I feel sure that at least part of the increase resulted from the information we gained through your team survey. Please accept our sincere appreciation for the efforts and extend our thanks to Sandy, Janee, and Elizabeth for their outstanding work in undertaking the survey.

Sincerely,

CEO

Dear (Professor):

I really appreciate your assistance to ________________. The customer survey was very helpful and the manager is considering all of your recommendations. The revenues from the store have helped place over a hundred disabled people in our community for the last five years. Your students' research will allow us to be more effective and productive.

Thank you very much.

President of the Board of Directors

Research can be a valuable tool in the change process, but by itself will probably not be sufficient to alter long-established attitudes and practices. The action research approach pioneered by Kurt Lewin (1946) described in Chapter 1 called for a combination of research and technical assistance in which the clients (potential users of the information) are intimately involved in all stages of gathering, analyzing, and disseminating the information. When people have been personally engaged in a study, they are more likely to understand the methods used in gathering the information and the implications of the findings. However, there is no sure and easy road to applying behavioral science findings. Changing organizations is as difficult (perhaps more so) as changing individuals. Several specialty areas such as organizational development and program evaluation have developed around the issue of organizational change.

News Releases

In addition to reaching other researchers through a scientific article, practitioners through a trade publication, and a client through a technical report, a researcher may also want to bring the findings to the attention of a wider audience, such as readers of local newspapers or the audience of local TV stations. This requires a different format and style than is appropriate for a journal article or technical report. Place the maximum emphasis on the findings and their implications for action, and eliminate any jargon. Even a modest study by a student researcher may rate a column in the local press if the issue is newsworthy. Such articles need not be a dull presentation of the results, boring the reader with endless statistics. A lively writing style can help maintain reader interest in the study without unduly distorting the findings.

There are dangers in premature media exposure. Not only can this disrupt a study in progress, but it also raises serious ethical questions if the findings contain errors. Science depends heavily upon *peer review,* which is the time-consuming process by which researchers critically evaluate one another's work before it is published. Scientists will be severely criticized by colleagues if they hold press conferences before their findings are reviewed by other researchers.

Talks at Meetings

Many professional organizations make it a policy to encourage student presentations and establish special sessions at their meetings for such papers. There are also student organizations that hold conferences to discuss research. This is excellent practice for subsequent talks at professional meetings. It can also provide useful feedback and is a nice addition to your application if you apply to graduate school. When making presentations at professional meetings, it helps to bear certain things in mind. Communication intended to be heard rather than read requires a different style. Assuming that you have already done a written report, you should prepare a separate outline for a talk. It is boring to hear someone read a written manuscript word for word. Since a major goal of presenting a talk to a live audience is to obtain feedback, be sure to leave time for questions. A few points well made will be more productive than trying to cram all of your study into a 15-minute session. Practice your talk on friends or colleagues beforehand. This will enable you to keep within the time limit and find out which statements are unclear.

Audiovisual Aids

Use visual aids wherever possible. Showing an audience what the experimental room or observational sheet looked like will have more impact than a verbal description. Things that people can see are likely to be remembered better than things merely talked about. Be sure that your displays or slides are visible and intelligible from all parts of the room. Know how to operate the audiovisual system prior to the session.

Frequent problems at talks include the following: previous session runs over and your talk starts late, faulty or incomplete introduction, microphone improperly adjusted or not working (e.g., emits high-pitched whistles), difficulty in locating proper switches for dimming lights, a slide in wrong order or upside down, other slides under- or overexposed making them too dark or too light, talk exceeds time limit). The best solution to these and other problems is to stage a dress rehearsal before a sympathetic audience. If possible, do it in the actual room where you will be speaking. If this is impossible, arrive early to familiarize yourself with the room and to check out the audiovisual equipment.

The ease of making transparencies for overhead projectors in copy machines has made this a very common method for presenting data at professional meetings. One can make transparencies of tables, charts, or illustrations directly from printed materials. However, don't get carried away and bombard your audience with complex tables of numbers. Select only the most important information for visual presentation and make sure in advance that it can be seen and understood by the audience in the time available. Turn off the projector when there is no need for it. This saves energy and reduces noise and distraction.

Summary

A single study points the way to other studies, leading to a research program involving several linked investigations. The review of papers by scientific journals is a time-consuming process. Potential authors need to exercise care in choosing from among available journals and have back-up plans in case the first journal rejects the paper. A manuscript is likely to go through several revisions (drafts) before it is finally accepted for publication.

A study may also have implications for practice. Practitioners can be reached through articles in occupational or trade periodicals. Action research stimulates change by involving the potential users of the information in the research process.

News releases and talks at meetings provide additional ways of getting your information to an audience. Keep such presentations lively and to the point using audiovisual aids when feasible. Practice your presentation ahead of time.

References

Lewin, K. (1946). Action research and minority problems. *Journal of Social Issues, 2,* 34–46.

Sommer, R., & Tyburczy, J. (1978, March). Waiting rooms: A study in user perception. *California Veterinarian, 32,* 21–23.

Further Reading

American Psychological Association. (1983). *Publication manual of the American Psychological Association* (3rd ed.). Washington, DC: Author.

Day, R. (1988). *How to write and publish a scientific paper* (3rd ed.). Philadelphia: ISI Press.

Morris, L. L., Fitz-Gibbon, C. T., & Freeman, M. E. (1987). *How to communicate evaluation findings.* Beverly Hills, CA: Sage.

Schlosberg, H. (1965). Hints on presenting a paper at an APA convention. *American Psychologist, 20,* 606–607.

Silverman, R. J. (1982). *Getting published in educational journals.* Springfield, IL: Charles C. Thomas.

Stone, J. & Bachner, J. (1977). *Speaking up: A book for every woman who wants to speak effectively.* New York: McGraw-Hill.

The *Journal of Applied Behavioral Science* and *Human Relations* emphasize articles on organizational change and the application of social science methods and findings.

22 Concluding Notes

This book began with an introduction to the multimethod approach. This was followed by a discussion of reviewing the literature and research ethics. We then moved to specific methods. First came observational methods, experimental design, interviews and questionnaires, followed by the less common techniques of content analysis, and the case study. We also described the data sources provided by personal documents and archives, as well as apparatus, standardized tests, calculators, and computers. Discussing each technique in a separate chapter may have undermined the notion of a multimethod approach. However, it seemed necessary to discuss each method separately, since the principles involved in running an experiment, for example, are different from those of doing an interview or analyzing personal documents. Having many techniques at your disposal is like being a skilled worker with a kit full of tools. In all likelihood, several tools rather than one will be needed on a specific job. Training and experience will aid you in their selection and use.

The statistical tests described in Chapters 17 and 18 are likely to seem abstract and difficult until you try to use them with your own data. When the questionnaires and observations have a personal significance because you yourself have collected them, the application of statistical tests will be more satisfying and meaningful than when you are using arbitrary numbers from a textbook example. Knowing the origins of statistical tests is not necessary in order to use them. Descriptive statistics are tools for reducing a mass of numbers to a few informative figures such as means, medians, and standard deviations. Inferential statistics are ways of analyzing data that allow you to choose among alternative explanations.

The chapters on reports attempted to place the research process within the larger context of science and practice. The study described was not earthshaking; few studies are. This one was undertaken by an undergraduate under the supervision of one of the authors. Portions of it were published—which indicates, among other things, that a study need not be profound in order to be published. As a beginning behavioral scientist, you should aim at doing good research—nothing more. A small, concise study properly done is a better learning experience for the beginner than a massive work that is never completed, analyzed, or written.

Before closing, we would like to offer a few additional suggestions to the beginning researcher. Some of these apply to specific techniques, others to behavioral research in general.

Helpful Hints

Always check reliability. No matter how simple and straightforward the behavior being observed, it is still necessary to check observer reliability. Remember the adaptation of Murphy's law—if something can be misinterpreted, it will be. The number of people in a room should be easy to record reliably. Not so—people come and go and move about. Even a count of billboards, road signs, or some other tangible items is likely to produce discrepancies between observers. Questions arise as to whether store signs along the road count as billboards, whether a sign divided into two parts is considered two signs or one, and so on. A scoring system considered reliable in earlier studies may be interpreted differently by a new observer.

Replication is useful. Don't worry if halfway through your research, you discover that someone has done a similar study. Science depends on repeated observations. If you come up with the same results, you have done a good job. If your results differ, this indicates a problem that must be resolved either through a finer-grained analysis or additional research.

Don't include too much in a single study. For a beginning researcher, there is a great temptation to keep on adding experimental conditions or test questions in the hope of positive results. There is the feeling of foreboding that the study will end inconclusively. Attempting to "cover yourself" by adding more conditions may create even further trouble. Every new variable increases the likelihood of finding relationships due to chance. If you compare enough things enough times, relationships will be found. Do not deceive yourself that such findings are meaningful. With unpredicted findings in a multivariable study, it is necessary to repeat the study at least one more time before there is any assurance that the obtained difference is not a random fluctuation.

Be careful about data collected by others. A tired interviewer may not be strongly motivated to do a good job. Occasionally interviewers have interviewed themselves and written the replies as if they came from other people. Pollster Albert Sindlinger declares, "If you hire somebody with any intelligence, after the tenth interview they are going to sit down and make up the other twenty-five" (Crouse, 1977). This view may seem a bit cynical, but it illustrates the need to check that field interviewers and observers are actually doing their job.

Specify in advance the characteristics for including subjects in the study as clearly and explicitly as possible (e.g., a good command of English, good eyesight and hearing with corrective devices, over age 18 and under 60, and so on). If possible, list the ways to identify nonserious answers in advance. One possibility is to discard replies that show a fixed or illogical pattern. There is reason to question a person who checks "strongly agree" to every statement, including those for and against an issue. Some researchers automatically discard any questionnaire that shows a stereotyped pattern of responses. Another way of determining nonserious answers is to ask the same questions several times in slightly different form. The person who strongly agrees with one form of the statement but disagrees with another is not a reliable respondent. It is useful to consult with colleagues, other researchers, or fellow students before discarding data. Occasionally, a response that looks unusual will make sense if you know the respondent's particular situation. Obtaining a second opinion will help answer the criticism that replies were rejected because they did not fit the hypothesis.

Double check information. Always check what you are told by those in authority. Seemingly factual statements such as "The park closes at 6:30 P.M." or "All the dormitory residents occupy double rooms" may be incorrect. It is surprising how often administrators are out of touch with what is going on in their own departments. The custodian is likely to know more about people's working habits than the company president. Don't be offensively distrustful of what people say or challenge their statements. Quietly check the facts and figures to see if they are accurate, complete, and current. Numerous horror stories could be recited of people who based their research projects on what authorities told them. A woman studying litter in city parking garages was told that the garages were swept every morning. Not so, she found, during several preliminary visits. The floors were cleaned once every 2 days at most. Had she not inspected the garages, her estimates of litter would have been grossly inflated.

Double check any maps or floor plans that are being used in an observational study. A faulty floor plan can invalidate otherwise good observations. In one study, observers recorded classroom participation in several lecture halls. It was later found that one observer had included an extra row of chairs and another had not indicated clearly the front of the room and occasionally reversed front and back locations. Neither of these errors was detected until after the study was completed. All their data had to be discarded.

Don't overanalyze the data. Only so much can be learned from a single study. Once the totals and averages have been computed, the major trends should be apparent. Overanalysis is a common error among beginning researchers, who are so eager to fully interpret the data that they tend to inflate the results. The amount of "noise" in behavioral research requires caution in interpreting trends that are not obvious. If 69 percent of the managers and 62 percent of the clerical employees are satisfied with working conditions, then the difference between them is probably not worth underscoring. The similarity between the two groups seems more significant than the difference.

When the major findings have been extracted from the data, you should seriously consider (1) stopping because you have learned what you wanted, or (2) undertaking another study. Either of these options is likely to be more sensible than spending the next 6 months attempting to squeeze the last bit of information out of the data. People become attached to their findings. This is quite normal; it is good to feel positive toward your findings. However, you should look upon each study as an attempt to answer certain questions. When this has been done, it is time to develop the implications of the findings and answer new questions that have arisen during the study. The danger of becoming bogged down in the data is particularly serious if your results have been analyzed by computer, creating a tremendous stack of printouts. You should accept the fact that most of the comparisons are probably not worth understanding or analyzing. Including them in the report will probably overwhelm the reader and detract attention from the major findings.

Not every study need be earthshaking. Science is a cumulative enterprise. Breakthrough discoveries occasionally occur, but most scientific investigation consists of small studies that shed light on one particular aspect of a topic.

Don't be concerned about the possibility that your methods or findings will be stolen. It is not necessary to copyright your term paper or class project. Sharing your findings with colleagues and fellow students can yield valuable feedback and ideas.

Don't be overly caught up in novelty. Nor does good research require demolishing all previous studies in the field. Behavioral science is based on a steady accumulation of solid studies that improves our understanding of human behavior. This includes small and large studies, minor and major adaptations of previous studies, new studies, and replications.

Avoid jargon in talking with outsiders. Scientific terms are intended for communication within the scientific community. A technical vocabulary is essential for keeping terms distinct and meaningful. Nonetheless, the same terminology that is helpful in communicating with other scientists may be a source of confusion and resentment when used with outsiders. Behavioral research is increasingly a collaborative effort, requiring communication with technical people from related fields and with lay people.

Be a responsible critic. Don't seize upon the limitations of a study in order to dismiss the entire project. Every study has its flaws. A responsible critic is obliged not only to indicate the nature of any limitations but how they operate to distort the data. Most of the landmark studies in behavioral research, including the classic investigations of Sigmund Freud, Jean Piaget, and Margaret Mead, were severely flawed in their methodology. The person who blazes a trail into a forest should not be criticized for not building a four-lane highway.

Be especially tolerant of the work of beginning researchers. They are the future of science. Try to specify what is good as well as what is missing in their studies.

Reference

Crouse, T. (1977, April). Washington column. *Esquire*, p. 16.

Appendixes

Appendix A Critical Values

TABLE A-1. Critical Values for t Test[a]

Degrees of freedom (df)	*Probability level*[b]	
	.05	.01
5	2.57	4.03
6	2.45	3.71
7	2.36	3.50
8	2.31	3.36
9	2.26	3.25
10	2.23	3.17
11	2.20	3.11
12	2.18	3.06
13	2.16	3.01
14	2.14	2.98
15	2.13	2.95
16	2.12	2.92
17	2.11	2.90
18	2.10	2.88
19	2.09	2.86
20	2.09	2.84
25	2.06	2.79
30	2.04	2.75
40	2.02	2.70
50	2.01	2.68
100	1.98	2.63
∞	1.96	2.58

[a]Abridged from Fisher and Yates, *Statistical tables for biological, agricultural, and medical research* (London: Longmans Group, Ltd.), 6th ed., 1974. (Previously published by Oliver and Boyd, Edinburgh.) Used with permission of the authors and publishers.

[b]These are *two-tailed* probability levels. They are used when the investigator has made no specific prediction about the direction of the difference. A *one-tailed* test is used when the investigator has predicted the direction of the difference. The probability levels for a one-tailed test are half those of a two-tailed test. Since two-tailed tests are more conservative, it is good practice to use them routinely.

TABLE A-2. Critical Values for Chi-square Test[a]

Degrees of freedom (df)	*Probability level*	
	.05	.01
1	3.84	6.64
2	5.99	9.21
3	7.81	11.34
4	9.49	13.28
5	11.07	15.09
6	12.59	16.81
7	14.07	18.48
8	15.51	20.09
9	16.92	21.67
10	18.31	23.21
11	19.68	24.73
12	21.03	26.22
13	22.36	27.69
14	23.68	29.14
15	25.00	30.58
20	31.41	37.57
30	43.77	50.89

[a]See footnotes for Table A-1.

TABLE A-3. Critical Values for Pearson Product-moment Correlation Coefficient (r)[a]

Degrees of freedom (df)	Probability level	
	0.5	.01
5	.754	.874
6	.707	.834
7	.666	.798
8	.632	.765
9	.602	.735
10	.576	.708
11	.553	.684
12	.532	.661
13	.514	.641
14	.497	.623
15	.482	.606
16	.468	.590
17	.456	.575
18	.444	.561
19	.433	.549
20	.423	.537
25	.381	.487
30	.349	.449
40	.304	.393
50	.273	.354
60	.250	.325
80	.217	.283
100	.195	.254
200	.138	.181
300	.113	.148
400	.098	.128

[a]See footnotes for Table A-1.

TABLE A-4. Critical Values for Spearman Rank-order Coefficient *(rho)*[a]

Number of pairs *N*	*Probability level*[b] .05	.01
5	.900	1.000
6	.829	.943
7	.714	.893
8	.643	.833
9	.600	.783
10	.564	.746
12	.506	.712
14	.456	.645
16	.425	.601
18	.399	.564
20	.377	.534
22	.359	.508
24	.343	.485
26	.329	.465
28	.317	.448
30	.306	.432

[a]Reprinted with permission from E. G. Olds, Distributions of sums of squares of rank differences for small numbers of individuals. *Annals of Mathematical Statistics,* 1938, 133–148.

[b]Unlike the earlier tables, which contain two-tailed probability levels, the levels for *rho* are one-tailed levels. They assume that the researcher has predicted the direction of the relationship as positive or negative. If no prediction has been made, the researcher can use the more conservative .01 probability levels.

TABLE A-5. Abridged Table of Random Numbers

73	17	86	15	27	10	42	72
64	46	07	88	76	45	61	55
19	09	22	09	35	99	76	34
08	72	60	22	84	17	81	39
90	09	97	61	90	37	23	52
34	43	09	17	30	20	59	61
02	48	34	21	18	03	99	61
45	90	33	88	89	70	04	80
03	61	05	61	77	70	17	47
05	56	83	78	26	48	35	06
15	86	60	14	49	10	51	17
38	07	45	88	06	06	29	92
60	22	86	92	52	31	00	47
81	94	25	53	73	89	42	62
87	97	01	09	03	40	86	12
17	35	11	60	12	23	83	26
71	27	96	45	07	60	71	82
66	38	80	72	74	42	21	53
94	84	69	37	69	35	59	32
03	26	07	66	93	88	48	54

TABLE A-6. Critical Values of F*

Degrees of freedom in denominator	*Degrees of freedom in numerator*													
	1	2	3	4	5	6	7	8	9	10	20	50	100	∞
1	161	200	216	225	230	234	237	239	241	242	248	252	253	254
	4052	**5000**	**5402**	**5625**	**5764**	**5859**	**5928**	**5980**	**6022**	**6056**	**6208**	**6302**	**6334**	**6366**
2	18.51	19.00	19.16	19.25	19.30	19.33	19.36	19.37	19.38	19.39	19.44	19.47	19.49	19.50
	98.50	**99.00**	**99.17**	**99.25**	**99.30**	**99.33**	**99.34**	**99.36**	**99.38**	**99.40**	**99.45**	**99.48**	**99.49**	**99.50**
3	10.13	9.55	9.28	9.12	9.01	8.94	8.88	8.84	8.81	8.78	8.66	8.58	8.56	8.53
	34.12	**30.81**	**29.46**	**28.71**	**28.24**	**27.91**	**27.67**	**27.49**	**27.34**	**27.23**	**26.69**	**26.30**	**26.23**	**26.12**
4	7.71	6.94	6.59	6.39	6.26	6.16	6.09	6.04	6.00	5.96	5.80	5.70	5.66	5.63
	21.20	**18.00**	**16.69**	**15.98**	**15.52**	**15.21**	**14.98**	**14.80**	**14.66**	**14.54**	**14.02**	**13.69**	**13.57**	**13.46**
5	6.61	5.79	5.41	5.19	5.05	4.95	4.88	4.82	4.78	4.74	4.56	4.44	4.40	4.36
	16.26	**13.27**	**12.06**	**11.39**	**10.97**	**10.67**	**10.45**	**10.27**	**10.15**	**10.05**	**9.55**	**9.24**	**9.13**	**9.02**
6	5.99	5.14	4.76	4.53	4.39	4.28	4.21	4.15	4.10	4.06	3.87	3.75	3.71	3.67
	13.74	**10.92**	**9.78**	**9.15**	**8.75**	**8.47**	**8.26**	**8.10**	**7.98**	**7.87**	**7.39**	**7.09**	**6.99**	**6.88**
7	5.59	4.74	4.35	4.12	3.97	3.87	3.79	3.73	3.68	3.63	3.44	3.32	3.28	3.23
	12.25	**9.55**	**8.45**	**7.85**	**7.46**	**7.19**	**7.00**	**6.84**	**6.71**	**6.62**	**6.15**	**5.85**	**5.75**	**5.65**
8	5.32	4.46	4.07	3.84	3.69	3.58	3.50	3.44	3.39	3.34	3.15	3.03	2.98	2.93
	11.26	**8.65**	**7.59**	**7.01**	**6.63**	**6.37**	**6.19**	**6.03**	**5.91**	**5.82**	**5.36**	**5.06**	**4.96**	**4.86**
9	5.12	4.26	3.86	3.63	3.48	3.37	3.29	3.23	3.18	3.13	2.93	2.80	2.76	2.71
	10.56	**8.02**	**6.99**	**6.42**	**6.06**	**5.80**	**5.62**	**5.47**	**5.35**	**5.26**	**4.80**	**4.51**	**4.41**	**4.31**
10	4.96	4.10	3.71	3.48	3.33	3.22	3.14	3.07	3.02	2.97	2.77	2.64	2.59	2.54
	10.04	**7.56**	**6.55**	**5.99**	**5.64**	**5.39**	**5.21**	**5.06**	**4.95**	**4.85**	**4.41**	**4.12**	**4.01**	**3.91**
11	4.84	3.98	3.59	3.36	3.20	3.09	3.01	2.95	2.90	2.86	2.65	2.50	2.45	2.40
	9.56	**7.20**	**6.22**	**5.67**	**5.32**	**5.07**	**4.88**	**4.74**	**4.63**	**4.54**	**4.10**	**3.80**	**3.70**	**3.60**
12	4.75	3.88	3.49	3.26	3.11	3.00	2.92	2.85	2.80	2.76	2.54	2.40	2.35	2.30
	9.33	**6.93**	**5.95**	**5.41**	**5.06**	**4.82**	**4.65**	**4.50**	**4.39**	**4.30**	**3.86**	**3.56**	**3.46**	**3.36**
13	4.67	3.80	3.41	3.18	3.02	2.92	2.84	2.77	2.72	2.67	2.46	2.32	2.26	2.21
	9.07	**6.70**	**5.74**	**5.20**	**4.86**	**4.62**	**4.44**	**4.30**	**4.19**	**4.10**	**3.67**	**3.37**	**3.27**	**3.16**

14	4.60	3.74	3.34	3.11	2.96	2.85	2.77	2.70	2.65	2.60	2.39	2.24	2.19	2.13
	8.86	**6.51**	**5.56**	**5.03**	**4.69**	**4.46**	**4.28**	**4.14**	**4.03**	**3.94**	**3.51**	**3.21**	**3.11**	**3.00**
15	4.54	3.68	3.29	3.06	2.90	2.79	2.70	2.64	2.59	2.55	2.33	2.18	2.12	2.07
	8.68	**6.36**	**5.42**	**4.89**	**4.56**	**4.32**	**4.14**	**4.00**	**3.89**	**3.80**	**3.36**	**3.07**	**2.97**	**2.87**
16	4.49	3.63	3.24	3.01	2.85	2.74	2.66	2.59	2.54	2.49	2.28	2.13	2.07	2.01
	8.53	**6.23**	**5.29**	**4.77**	**4.44**	**4.20**	**4.03**	**3.89**	**3.78**	**3.69**	**3.25**	**2.96**	**2.86**	**2.75**
17	4.45	3.59	3.20	2.96	2.81	2.70	2.62	2.55	2.50	2.45	2.23	2.08	2.02	1.96
	8.40	**6.11**	**5.18**	**4.67**	**4.34**	**4.10**	**3.93**	**3.79**	**3.68**	**3.59**	**3.16**	**2.86**	**2.76**	**2.65**
18	4.41	3.55	3.16	2.93	2.77	2.66	2.58	2.51	2.46	2.41	2.19	2.04	1.98	1.92
	8.28	**6.01**	**5.09**	**4.58**	**4.25**	**4.01**	**3.85**	**3.71**	**3.60**	**3.51**	**3.07**	**2.78**	**2.68**	**2.57**
19	4.38	3.52	3.13	2.90	2.74	2.63	2.55	2.48	2.43	2.38	2.15	2.00	1.94	1.88
	8.18	**5.93**	**5.01**	**4.50**	**4.17**	**3.94**	**3.77**	**3.63**	**3.52**	**3.43**	**3.00**	**2.70**	**2.60**	**2.49**
20	4.35	3.49	3.10	2.87	2.71	2.60	2.52	2.45	2.40	2.35	2.12	1.96	1.90	1.84
	8.10	**5.85**	**4.94**	**4.43**	**4.10**	**3.87**	**3.71**	**3.56**	**3.45**	**3.37**	**2.94**	**2.63**	**2.53**	**2.42**
25	4.24	3.38	2.99	2.76	2.60	2.49	2.41	2.34	2.28	2.24	2.00	1.84	1.77	1.71
	7.77	**5.57**	**4.68**	**4.18**	**3.86**	**3.63**	**3.46**	**3.32**	**3.21**	**3.13**	**2.70**	**2.40**	**2.29**	**2.17**
30	4.17	3.32	2.92	2.69	2.53	2.42	2.34	2.27	2.21	2.16	1.93	1.76	1.69	1.62
	7.56	**5.39**	**4.51**	**4.02**	**3.70**	**3.47**	**3.30**	**3.17**	**3.06**	**2.98**	**2.55**	**2.24**	**2.13**	**2.01**
40	4.08	3.23	2.84	2.61	2.45	2.34	2.25	2.18	2.12	2.07	1.84	1.66	1.59	1.51
	7.31	**5.18**	**4.31**	**3.83**	**3.51**	**3.29**	**3.12**	**2.99**	**2.88**	**2.80**	**2.37**	**2.05**	**1.94**	**1.81**
50	4.03	3.18	2.79	2.56	2.40	2.29	2.20	2.13	2.07	2.02	1.78	1.60	1.52	1.44
	7.17	**5.06**	**4.20**	**3.72**	**3.41**	**3.18**	**3.02**	**2.88**	**2.78**	**2.70**	**2.26**	**1.94**	**1.82**	**1.68**
100	3.94	3.09	2.70	2.46	2.30	2.19	2.10	2.03	1.97	1.92	1.68	1.48	1.39	1.28
	6.90	**4.82**	**3.98**	**3.51**	**3.20**	**2.99**	**2.82**	**2.69**	**2.59**	**2.51**	**2.06**	**1.73**	**1.59**	**1.43**
200	3.89	3.04	2.65	2.41	2.26	2.14	2.05	1.98	1.92	1.87	1.62	1.42	1.32	1.19
	6.76	**4.71**	**3.88**	**3.41**	**3.11**	**2.90**	**2.73**	**2.60**	**2.50**	**2.41**	**1.97**	**1.62**	**1.48**	**1.28**
∞	3.84	3.00	2.60	2.37	2.21	2.10	2.01	1.94	1.88	1.83	1.57	1.35	1.24	1.00
	6.64	**4.61**	**3.78**	**3.32**	**3.02**	**2.80**	**2.64**	**2.51**	**2.41**	**2.32**	**1.87**	**1.52**	**1.36**	**1.00**

*The F ratio is statistically significant when it exceeds or equals the values above. The .05 level is shown in light type; the .01 level in bold type.

Appendix B Example of a Manuscript Prepared According to the Style Manual of the American Psychological Association

The following study was a multimethod collaboration between two authors at different universities. One author conducted the semantic differential studies and the other author did the reaction time study using different samples. This version of the paper has been adapted and rewritten according to the style manual of the American Psychological Association (APA). Had the article been written for a non-APA journal, a different style might have been required. When you write an article for a specific periodical, you should consult the style guide issued by the publisher. If you are writing a research report not intended for publication, check the style used by others who have written similar reports.

Perception of Traffic Signs Containing Conflicting Symbol and Direction Information

Leslie A. Whitaker

University of Missouri - St. Louis

Robert Sommer

University of California, Davis

Running Head: TRAFFIC SIGNS

Abstract

Road information signs (airport and ski trail symbols), each with directional quality, were paired with route guidance arrows. The combinations were rated on semantic differential scales. Response times to the combination signs and arrows were measured in a second sample. Concordant signs, for which the symbol and arrow agree, were judged to be more clear, unified, valuable, and strong than signs containing discordant symbols and arrows. Response times were slower for the discordant airport signs than for the comparable concordant signs. However, concordance did not influence response time to the skier. It was concluded that agreement between symbol and arrow direction is an important element in reducing perceptual conflict of road signs.

PERCEPTION OF TRAFFIC SIGNS CONTAINING CONFLICTING SYMBOL AND DIRECTION INFORMATION

Current designs for traffic signs emphasize a conversion from printed words to pictorial symbols. Pictorial symbols have the obvious advantage of bypassing the need for two languages on road signs in bilingual cultures and of improving the ease with which travelers can move through an unfamiliar country for business or pleasure (Mackett-Sloat and Dewar, 1981). In addition, pictorial symbols are legible to drivers at greater distances than printed messages (Dewar, Ells, and Mundy, 1976), are more easily recognized when presented in a degraded form (Jacobs, Johnston, and Cole, 1975) and have better glance legibility than word signs (Walker, Nicolay, and Stearns, 1965).

In general, road guidance signs consist of a pictograph combined with an arrow which conveys the direction to the object. Since some pictographs have strong directional information portrayed in their shape (e.g., an airplane pictograph), there is the possibility of conflict between the symbol and the correct direction. When a motorist sees an airplane symbol flying to the right and the arrow pointing to the left, does this create a perceptual conflict? The motorist may not be able to ignore the direction cues inherent in the swept wing jet symbol (Figure 1a). This would result in an interference effect which would decrease the time necessary to process the directional arrow information (Dewar, 1976; Stroop, 1935). When the driver must respond quickly and accurately to a guidance sign, any conflict between the implied direction of the graphic symbol and that of the route arrow will have adverse consequences on comprehension and response time. The issue has much broader significance than this single traffic symbol. Traffic guidance symbols for railroads, subways, ferries, and tourist areas employ graphic symbols with directional quality.

Insert Figure 1 about here

Two types of perceptual conflict found on existing road signs were identified. The first was a discordant combination in which pictograph and arrow faced opposite directions (Figure 1b) and the second type of conflict, an orthogonal combination in which the direction of these two elements varied by 90 degrees (Figure 1c). The hypothesis tested in the present study are that (a) concordant signs in which a pictograph and arrow face in the same direction (e.g., Figure 1a) will be rated to have less perceptual conflict and will be responded to more quickly than either orthogonal or discordant signs and (b) orthogonal signs will be rated to have less perceptual conflict and will be responded to more quickly than discordant signs.

Experiment 1: Semantic Differential Ratings

Method

Subjects. Students at the University of California, Davis, participated as subjects. All were licensed drivers between the ages of 18 and 26. All 61 students in one class rated the airplane figures while all 14 students in a second class rated the skier figures.

Stimuli. The airport symbol and the cross-country skier symbol were the two symbols tested in this experiment. The pictographs for these stimuli were drawn in black ink on a simulated highway traffic sign. The airport symbol was oriented up, down, left or right. When paired with the three directional arrows, this produced 12 airport signs. The skier symbol could be oriented only right or left, producing a total of 6 skier signs. Each of these 18 stimuli was reproduced as a black and white slide.

<u>Procedure</u>

Each class was tested as a group. Subjects were told that they would see a series of signs containing a symbol and a directional arrow. They were to rate each sign on six adjective (semantic differential) scales. Rating forms were provided. The order of the scales was counterbalanced across subjects and stimuli; the positive end of the scale was reversed for half the cases.

Six semantic differential scales were used. These included three directed specifically to the experience of perceptual conflict (clarity, unity, and tension), and three expressing the major dimension of meaning (value, strength, activity) (Osgood, Suci, and

Tannenbaum, 1957). Five-point rating scales were used.

Results

The prediction had been made that the concordant combinations would be rated as superior to the discordant orthogonal sign combinations, and the orthogonal (right angle) combinations would be rated as superior to the discordant combinations on all dimensions. Table 1 shows the mean ratings of the airport symbol on the six semantic differential scales. ANOVA tests confirmed that the predictions were sustained at a very high level of confidence ($p < .01$) on all rating scales.

Subsequent analysis by t-tests showed significant differences between the sign combinations on all six dimensions, except on the activity dimension where there was no difference between the orthogonal and the discordant signs.

Insert Table 1 about here

Table 2 shows significant differences in all dimensions in semantic differential ratings of the cross-country skier sign combinations. Subsequent *t*-tests showed no difference between concordant and orthogonal combinations on the tension scale and no difference between the orthogonal and discordant signs on the strength and activity scales.

Insert Table 2 about here

Experiment 2: Reaction Time

Method

Subjects. Thirty students from introductory psychology classes at the University of Missouri-St. Louis participated as subjects in a reaction time experiment. Half the subjects were women. Each subject was tested individually. All subjects had 20/25 or better vision when tested on a Snellen chart.

Stimuli. The same 18 signs used in Experiment 1 were reduced to 3.5 x 5.5 cm. Five copies of each sign were made. Each of the resulting 90 stimuli was placed on separate cards as follows: one copy of each sign was placed in each of the four corners and at the center of the tachistoscopic viewing field. These locations were chosen to prevent the subject from fixating on a single location of the arrow and ignoring the pictograph symbol.

Procedure. Each subject was tested in a single session. The 90 stimulus cards were divided into five blocks of trials. Each block contained one instance of each sign. The locations of these 18 signs were divided as equally as possible among the five possible positions on the card. Stimuli were presented in a two-channel Gerbrands Model 125 tachistoscope.

The subject initiated each trial with a button press. After a two-second warning interval, the stimulus appeared for 200 ms. The subject responded as quickly as possible by saying the direction of the arrow, as "left," "right," or "up." The response closed a voice-operated relay (Gerbrands' Model G1342). The experimenter recorded the response latency from the digital readout of a Hunter Klockounter (Model 120C). Subjects were first trained on a separate block of training trials using a similar set of the 18 signs. They then completed the five reaction time blocks. If a subject made an error, the stimulus was retested later within the same block of trials.

Results

Airport. Table 3 shows that the concordant conditions resulted in the fastest responses, while the orthogonal conditions were slower, and the discordant conditions were the slowest. The overall effect was significant by ANOVA test, $F(3,84) = 22.95$, $p < .001$.

Insert Table 3 about here

Skier. Arrow direction did not have a significant effect on response time for this figure, $F(1,28) = 1.18$, p = ns). The discordant conditions resulted in a lower response of only 16 milliseconds, from 833 ms to 849 ms.

Discussion

The hypotheses of this study regarding semantic differential ratings of the various sign combinations were confirmed at a very high level of confidence. The concordant sign combinations were more clear, unified, valuable and strong than the conflicting signs. The only scale on which there was minimal or little difference was in the activity dimension. It seems likely that the perceptual conflict created a dynamic tension which the respondents may have interpreted as activity.

Reaction times to these symbols showed a large effect of concordance for the airport signs, i.e., those with strong directional symbols. In contrast, concordance produced no effect on response time with the skier symbols. These have weaker directional information since the skier's strongest line is a positive diagonal. Although the skier is facing in a specific direction, this symbol (unlike the airplane) could not easily be confused with an arrow. This may explain the fact that subjects rated the discordant sign as less clear, strong, and valuable on the semantic differential but were still able to process the guidance arrow as quickly as with concordant signs.

The combination of perceptual conflict ratings and reaction times is consistent with Kahneman's (1973) model of the Stroop phenomenon. He proposed that perceptual processing of various stimulus features proceeds as a race to their respective responses. When two or more stimulus features prompt conflicting responses, errors and total processing time increase. The traffic signs tested in the present study prompted two responses, one to the directional quality of the symbol and the second to the guidance arrow. In the case of discordant signs, these two responses are bipolar opposites (left versus right, up versus down); therefore, the greatest conflict occurs because two incompatible responses in the same perceptual dimension have been prompted. The conflict between competing responses to the orthogonal signs is less extreme because they are on two separate directional dimensions (vertical and horizontal).

Some government agencies have recognized the potential conflict between traffic logos and directional arrows. The Australian standard for the design of graphic devices states explicitly that symbol design shall avoid any elements that suggest directional information, unless the symbol is specifically intended to convey a direction (Standard Association of Australia, 1981). The reaction-time data in the present studies support this policy. Concordant airport signs required considerably less processing time than discordant ones. Conflicting directional cues associated with transport logos produced delays in reaction time in a quick-response situation. This is a factor if motorists under pressure of city traffic have only brief intervals to scan and process information regarding the correct route to the airport or railway station.

While some agencies have adopted policies regulating symbols that contain directional quality that may conflict with traffic directions, other agencies have not. Once a particular symbol is decided upon and installed, it will be difficult to have it removed or changed. The more logical approach is the prevention of perceptual conflict through adequate pre-testing of designs rather than attempting to remove or alter signs that have been adopted. As Ballinger and Ballinger (1972) indicated, the message on directional signs "must be visual, direct, and easily remembered so that immediate action can be taken" (p. 170).

References

Ballinger, L. B., & Ballinger, R. A. (1972). Sign, symbol, and form. New York: Van Nostrand Reinhold.

Corballism M., & Beale, C. (1976). The psychology of left and right. New York: Wiley.

Dewar, R. E. (1976). The slash obscures the symbol on prohibitive traffic signs. Human Factors, 18, 253-258.

Jacobs, R. J., Johnston, A. W., & Cole, B. L. (1975). The visibility of alphabetic and symbolic traffic signs. Australian Road Research, 5(7), 68-86.

Kahneman, D. (1973). Attention and effort. Englewood Cliffs, NJ: Prentice Hall.

Mackett-Sloat, J., & Dewar, R. (1981). Evaluation of symbolic public information signs. Human Factors, 23, 139-151.

Osgood, C. E., Suci, G., & Tannenbaum, P. (1957). The measurement of meaning. Urbana, IL: University of Illinois Press.

Standard Association of Australia (1981). Design and use of graphic symbols and public information symbol signs. Part 4: Principle for the design of graphic symbols. Canberra, Australian Printing Office.

Stroop, J. R. (1935). Studies of interference in verbal reactions. Journal of Experimental Psychology, 18, 643-663.

Walker, R. E., Nicolay, R. C., & Stearns, C. R. (1965). Comparative accuracy of recognizing American and international road signs. Journal of Applied Psychology, 49, 322-325.

Table 1

Ratings of Airport and Arrow Combinations

Adjective scale	Mean rating			F
	Concordant	Conflicting	Discordant	
unclear-clear	4.15	1.87	2.80	54.7*
weak-strong	3.95	1.92	2.52	63.6*
valuable-worthless	3.92	2.07	2.76	39.0*
active-passive	3.73	3.03	2.98	11.0*
tense-relaxed	3.59	1.75	2.40	51.9*
unified-separate	4.35	1.54	2.40	112.9*

*$p < .01$

Table 2

Ratings of Skier and Arrow Combinations

Adjective scale	Mean rating			
	Concordant	Conflicting	Discordant	*F*
unclear-clear	4.29	2.15	3.21	8.9**
weak-strong	3.90	2.43	2.86	5.2**
valuable-worthless	2.00	3.93	2.86	13.4**
active-passive	2.10	2.25	2.47	2.7*
tense-relaxed	2.84	2.14	2.75	4.3**
unified-separate	1.82	4.14	3.00	15.3*

*$p < .05$ **$p < .01$

Table 3

Reaction Time for Airport Sign Combinations

Airport-Arrow combination	Mean Response (milliseconds)
Concordant	795
Discordant	870
Orthogonal	837

a. Concordant b. Discordant c. Orthogonal

FIG. B-1. Combinations of symbols and directional arrows.

Appendix C Example of the Method Section in an Observational Study Written According to the Style Manual of the American Psychological Association

The study of traffic signs described in Appendix B involved an experiment. For various reasons, experiments are easier to package in APA style format than other types of studies. We have found that students often experience difficulty in fitting observational studies into an APA format. The problem is most apparent in the Method section. The following example shows the Method section of the coffeehouse observations included in Chapter 4 written in APA style. Instead of describing the apparatus, there is a paragraph on the settings where the observations took place.

Method

Settings. The three coffeehouses, all located in a university city with a population of 36,000, were different in many respects. Coffeehouse 1 was located in the downtown business district and appealed primarily to office employees and merchants on their coffee breaks. Coffeehouse 2 was located in a shopping center abutting a residential neighborhood and attracted primarily shoppers. The third coffeehouse was located in the Student Union building of the university and was patronized almost exclusively by students and university employees. Density was moderate in the downtown coffeehouse, low in the shopping center facility, and high in the student union.

Observations. The observers (Os) were university students who recorded their observations on prepared forms with categories for age in 10-year intervals, gender, group status (lone, group, or joined), duration of stay, food and beverage consumption, and reading. These categories were found to be capable of reliable scoring by student Os in an earlier investigation. All observations were made unobtrusively by Os who appeared to be normal participants in the setting.

The three coffeehouses were observed during successive quarters by different Os. These were considered to be three independent tests of the hypothesis, and they are referred to as Studies 1, 2, and 3, respectively. The number of observation sessions, which usually lasted from 60 to 120 minutes, was 23, 12, and 17, respectively, in the three studies. All observations took place during *non*meal hours when there would be empty seats and minimal pressure to finish up quickly and leave.

For each session, O entered the coffeehouse, purchased a beverage, and sat at a table that provided a good vantage point. The O selected the next three lone individuals and three individuals in groups entering the premises who sat down within O's field of vision. Each session included a *maximum* of three lone and three group drinkers. This limit was intended to minimize the potential impact of any single session on the final results. People entering the coffeehouse to purchase take-out pastries who did not sit down at tables were not included in the sample. In groups, O randomly chose one individual for observation and ignored other individuals. During some sessions it was not possible to fill the quota, and fewer than three lone or group drinkers were observed. Individuals were classified as lone or group patrons at the outset, and, if they were joined later, this was recorded.

Appendix D Example of Style Used for a Technical Report

The following report was produced in England by a research laboratory affiliated with a government agency. It is a good example of a brief, focused, and well-written technical report. The problem, methods, findings, and conclusions are clearly stated. The limitations of the study, in terms of artificiality of the experimental roadway and the experienced volunteer subjects, are mentioned. The full document consisted of a blue-and-white cover of a style used by the research laboratory, 4 pages of introductory material, 17 pages of text, and 12 pages of appendices. Because the report is used in this section to illustrate style, some of the details are abridged or omitted. The report is presented through the courtesy of the Transport and Road Research Laboratory, Berkshire, England.

TRANSPORT AND ROAD RESEARCH LABORATORY
Department of the Environment
Department of Transport
SUPPLEMENTARY REPORT 357

ATTITUDES OF A SAMPLE OF CYCLISTS TO USING SINGLE-LANE ROADS

F. J. Nicholson

1978

Table of Contents

ATTITUDES OF A SAMPLE OF CYCLISTS TO USING SINGLE-LANE ROADS

Abstract

This report describes an experiment to determine the attitudes of 25 cyclists to sharing single-lane roads with motor vehicles. Three widths of road (2.75 m, 3.00 m and 3.25 m) were examined and, on each width of road, all cyclists both passed a car traveling in the opposite direction and were overtaken by a car.

Objective measurements and the responses given by the cyclists to questionnaires are presented and compared. Variations in results according to road width, age of cyclist and manoeuvre undertaken are discussed.

Generally, no significant difference was found between results for the two age groups of cyclists (under sixteen and sixteen or over) or for the two manoeuvres. Cyclists stated that they felt unsafe in 22 of the 50 runs on the 2.75 m wide road, in 10 of the 49 runs on the 3.00 m wide road and in 0 of the 49 runs on the 3.25 m wide road. Whilst there was a diversity of opinion amongst the cyclists over whether single-lane roads are a good concept, a width of 3.00 m for such roads was found to be acceptable to many subjects and 3.25 m to virtually all.

No attempt is made in this report to predict the likely safety record of such single-lane roads.

1. Introduction

The Housing Development Directorate (HDD) of the Department of Environment has proposed that single-lane access roads up to 4.00 m in width could be used on housing estates. Guidelines on the design of such roads have been published recently as a Design Bulletin. The Laboratory has cooperated with HDD in investigating various aspects of this proposal. One investigation was concerned with the use of such single-lane roads by cyclists. This Report describes an experiment conducted at TRRL in August 1976 to determine the attitudes of a number of cyclists towards sharing single-lane roads with motor vehicles. No attempt is made to predict the likely safety record of such single-lane roads.

Recommendations are given on the minimum width required for single-lane roads if cyclists are to find them acceptable.

2. Experimental Method

2.1 Experimental Layout

. . . In planning the cycling experiment, it was decided that it was necessary to determine the attitude of cyclists to the "worst" conditions that they would be likely to encounter regularly on the access roads. It was therefore decided to investigate the attitudes of the cyclists to two manoeuvres that they would have to execute on single-lane roads. The manoeuvres were (a) to meet and pass a motor vehicle at a point between two passing bays, and (b) to be overtaken by a motor vehicle at a point between two passing bays . . .

2.2. Experimental Procedure

Two tests per cyclist were carried out for each of the three experimental widths . . .

2.3 Selection of Cyclists for the Experiment

Cyclist subjects were obtained by placing advertisements in local papers, asking for volunteer cyclists aged 16 and over, and by issuing a circular within the laboratory asking for volunteers aged under 16 from among children of staff . . . The overall division of the subjects by sex and by eligibility to hold a motor vehicle license corresponded with the distribution of an earlier travel survey . . .

2.4 Selection of Drivers and Vehicles for the Experiment

As school children were involved in the experiment, vehicles had to be driven by (drivers associated with the laboratory). Four drivers took part in the experiment. Careful consideration was given to the type of motor vehicle to be used . . .

2.5 Method of Analysis

The analysis of the experiment fell into three sections (a) analysis of the time-lapse film, (b) analysis of the questionnaire responses, and (c) comparison of the result of the film and questionnaire analyses . . .

3. Results

(This section contained 11 pages, including nine tables with descriptive text.)

4. Conclusions

1. The conditions for the experiment and the experience and physical stability of the 25 cyclists taking part were, on the average, better than might be expected in reality . . .
2. The results of the experiment did not, in general, differ according to either the age group of the subjects or the type of manoeuvres carried out (meeting, passing, or overtaking).
3. Cyclists said they felt unsafe in 22 of the 50 runs simulating a road 2.57 meters wide, in 10 of the 49 runs on a 3.00 meter road, and on none of the 49 runs on a 3.25 meter road . . .
4. Thirteen of the 25 subjects at some time thought that other cyclists would have felt less safe. . . . The main reasons given by the subjects for feeling safer were that they considered themselves to be more experienced and more physically stable than other cyclists and that the experimental conditions were artificially safe.
5. Observations from film of the experiment showed that the incidence of motorists driving off the road decreased as the width of the road increased. . . .
6. Cyclists generally failed to report significant events such as their slowing down or motorists driving off the road. They were also poor judges of whether the motorist had left as much clearance as possible. . . .
7. There was a diversity of opinion among the cyclists over whether the sharing of single-lane roads with motor vehicles is a good idea.
8. Given the limitations of conclusions 1–7 above, an actual road width of 3 meters is acceptable to a substantial number of the subjects and an actual road width of 3.25 meters is acceptable to virtually all subjects.
9. The results of this experiment show clearly the attitudes of a small number of cyclists to the shared use of single lanes but cannot give any guide to the likely safety record of such facilities.

5. Acknowledgments

(Statement of thanks to cooperating agencies.)

6. References

(Presented in numerical order as they appeared in text.)

7. Appendix

(Copies of questionnaires, diagrams and photograph of experimental roadway, and bar graphs showing results.)

Glossary

Accidental Sample A nonprobability sample comprised of those individuals available for study. Sometimes called a convenience sample.

Accretion The buildup of a residue or product of an interaction; something added to the situation.

Action Research Combines the testing of theory with application. Approach pioneered by Kurt Lewin.

Aggregate Data Analysis Studying the characteristics of groups using census data or other social statistics.

Alternative Hypothesis Although the null hypothesis is stated in negative terms (that there will be no effect), the working or alternative hypothesis is stated in positive terms (i.e., that there will be an effect).

Analysis of Covariance An adaptation of analysis of variance that takes into account scores on an additional variable believed to affect the independent variable.

Analysis of Variance (ANOVA) Statistical technique for comparing differences between means. See Chapter 18 for computational procedures.

Anonymity Refers to a situation in behavioral science research in which the researcher does not know the identity of the participants in the study (i.e., observing people at a street corner or bus station).

ANOVA See *Analysis of Variance.*

Applied Research Uses systematic procedures to provide answers to pressing questions. Motivated primarily by the need to solve an immediate problem.

Archives Public records and documents.

Attitude Scale A type of questionnaire designed to produce scores indicating the overall degree of favorability of a person's attitude on a topic.

Average A generic term referring to various measures of central tendency—the mean, median, and mode.

Balance In survey research, balance refers to the neutrality of questions and providing sufficient items so that those leaning toward one view are balanced by an equal number leaning toward the other view.

Bar Graph Graph made up of bars showing frequency of occurrence of scores or category levels. Also called a histogram.

Basic Research Investigation designed to answer general long-range questions. Motivated primarily by curiosity or testing theory.

Behavioral Map A chart of people's locations in space. Maps can be place-centered or person-centered.

Behavior Setting Geographic location linked to customary patterns of behavior.

Bibliography A listing of titles and citation information of works on a topic.

Blind Testing In sensory evaluation or drug research, blind testing indicates that the subject is not aware of the identity of the stimulus. If the experimenter is also unaware of the identity of the stimulus, this is called a *double blind procedure.*

Case Study In-depth investigation of a single instance. The unit can be as small as an individual or as large as an entire community.

Casual Observation Observation without prearranged categories or scoring system. Eyeball inspection of what is happening.

Categorical Variable Variable with mutually exclusive categories, whose levels are more qualitative than quantitative. Gender and place of residence are generally categorical variables.

Chi-square A statistical test used with categorical data to test whether an obtained distribution of scores differs reliably from what would be expected by chance.

Closed Questions Respondents are asked to choose among fixed alternative answers; also known as multiple-choice questions.

Codebook A listing of variable names and value labels used in a computer analysis.

Coding The transformation or reduction of raw data into a set of standard categories for statistical analysis. Frequently used in scoring responses to open-ended interview questions and in content analysis of mass media.

Cognitive Map A cognitive representation of the location of objects in space.

Concurrent Validity The correlation of a test with present behavior, or with other existing tests or measures; a subtype of criterion validity.

Confidentiality Participant or respondent identity is known to researcher but is not publicly revealed. Frequently achieved by removing identifying information from published reports.

Confounding Confusion of the effects of variables, resulting in an inability to determine which variable is the cause of an observed effect.

Connotative Meaning The subjective or personal meaning of something, as distinct from its physical characteristics.

Construct Validity The linkage of the test measurement to specific theoretical constructs, the relationship of test to theory.

Content Subject matter; in content analysis, refers to specific topics or themes, in contrast with format that refers to form or structure.

Content Analysis A technique for systematically describing the form and content of written or spoken material. Frequently used for quantitative study of mass media.

Content Validity The degree to which test items assess the domain that a test claims to cover, the relevance of the items to the behavior measured.

Contingency Table A table used to record the relationship between two or more variables. The observed frequencies are placed into the cells of the table.

Continuous Variable Variable with a continuous distribution of scores, such as age or income.

Control Group A subject group that resembles the experimental group in every respect except that it is not exposed to the independent variable. Used to control for the effects of extraneous variables on the dependent variable.

Converging Operations Using several procedures rather than one in conducting research on a topic.

Correlation The association between two sets of scores. Often expressed in terms of a *correlation coefficient,* which is a numerical indication of the degree of relationship between two sets of scores.

Counterbalancing Reversing the order of presentation of variables.

Criterion Validity The relationship of the test scores to other measures of the characteristics.

Cross-sectional Study A comparison of different groups at a single time.

Cumulative Scores Accumulation of frequencies from the bottom to the top of a frequency distribution, formed by starting at the bottom of a frequency distribution and summing all the frequencies at and below each level. Can be transferred into a *cumulative percentage* by dividing each score by the total number of cases and multiplying by 100.

Data List A printout of the computer file containing a complete data set.

Degrees of Freedom The number of values that are free to vary after certain restrictions have been placed on the data. Used to evaluate the results of various statistical tests.

Dependent Variable Assumed to be influenced by or dependent on the independent variable. If you study the effects of alcohol on speed of response, then speed of response is the dependent variable in this experiment.

Depth Interview A form of unstructured interview in which the interviewer follows the respondent's answers with a request for more information at an increasing level of depth.

Descriptive Statistics Techniques to organize and summarize data. Numerical values describing the characteristics of or within a sample.

Double Blind Procedure A research design in which neither the experimenter nor subject is aware of the treatment condition to which the subject has been assigned.

Equivalent Forms Two different, but comparable forms of a scale; used for assessing reliability or to avoid practice effects.

Erosion The deterioration or wear that provides an index of usage patterns; something is worn down or removed.

Ethnography A type of qualitative research involving the description and study of specific peoples and places.

Experimental Group A subject group exposed to the independent variable; sometimes called the treatment group.

External Validity The generalizability of research findings.

Extraneous Variable A source of error, other than the independent variable, that might influence the dependent variable.

Face Validity The appearance of being a valid measure of something; a measure that "looks right" to an outside observer.

Factor The independent variable or treatment. A study investigating the effects of age and education on attitude would be a two-factor study. If additional factors, such as gender were added, this would become a multifactorial study.

Factorial Design Involves more than a single factor or independent variable in a study.

Field Experiment Experiment in which the independent variable is manipulated in a natural setting rather than in a laboratory.

Format Structure; pertains to the appearance, order, and wording of items in a questionnaire or interview; in contrast to content, which refers to meaning.

Frequency Number of times a score or category level occurs.

Frequency Distribution Arrangement of scores from highest to lowest, together with the frequency of each score (i.e., the number of times each score occurs).

Frequency Polygon or Curve A type of graph representing a frequency distribution, in which a continuous line is used to indicate the frequency of the corresponding scores.

Graphic Rating Scale The respondent places a mark along a continuous line.
Halo Effect The tendency to make ratings of specific abilities on the basis of an overall impression.
Hard Disk A piece of computer hardware that can store a large amount of information permitting complex data processing.
Hardware The machine parts of a computer.
Histogram Similar to a bar graph.
Horizontal Axis The straight line at the base of a graph.
Hypothesis A research question phrased in the form of a statement; a testable proposition.
Independent Variable The variable that is systematically manipulated by the researcher. Also known as the experimental variable.
Inferential Statistics Statistical techniques used to make generalizations from samples to populations.
Institutional Review Board A group of people with formal responsibility for reviewing submitted research proposals in terms of ethics and protection of the participants.
Instrumental Research Undertaken as an academic, vocational, or professional requirement. Goal is to demonstrate competence in research.
Internal Validity In an experiment, the degree to which the experiment does not have confounding extraneous variables; the degree to which the independent variable did indeed produce changes in the dependent variable.
Inventory See *test.*
Latent Content The less obvious and hidden information, the underlying meaning of a response as distinct from the manifest content, which is on the surface and obvious.
Legend Title and brief description accompanying a table, figure, or graph.
Level The amount or value of a variable; dimensions along which a single variable differs. Using hair color as a variable, the levels are blond, brunette, etc.
Library of Congress Subject Heading Index A listing of topics under which books are catalogued in libraries.
Life History Biographical technique useful in qualitative research.
Likert Scale A type of attitude scale containing statements that are clearly favorable or clearly unfavorable. The respondent indicates degree of agreement.
Longitudinal Study The study of an individual or group over a long time period.
Mainframe Computers Computers capable of manipulating large data sets, generally available only at large institutions.
Manifest Content That portion of an interview or response that is obvious and contained in the spoken or written information. Can be contrasted to latent content, which is less obvious.
Matched Groups Assigning subjects so that the experimental and control groups are as similar as possible. A means of controlling extraneous variables.
Mean Arithmetic average; sum of scores divided by the number of cases.
Median Midpoint of the distribution when all the scores are arranged from highest to lowest. Half the scores fall above the median and half below.
Microcomputers Small, desk-size personal computers.
Mode The single score that occurs most often in a distribution. If there are two modes, the distribution is called *bimodal.*
Multivariate Analysis Analysis of the interrelationships among several variables at

the same time. For example, in an opinion survey, the effects of age, education, and income could be analyzed simultaneously on more than one outcome through the use of multivariate analysis.

Natural Experiment A study that uses a naturally occurring event or policy change as the independent variable. Not a true experiment because the researcher does not directly manipulate or control the independent variable, and/or the subjects are not randomly assigned to treatment conditions.

Networking A strategy of gaining information by using the references in recent articles to find other related articles or information.

Nonprobability Samples The likelihood of selection is not actually known to the researcher. Three general types are quota, purposive, and accidental samples.

Normal Curve A symmetrical bell-shaped curve which often approximates the frequency of occurrence of events in nature.

Norms Statistical summary of people's performance on tests or inventories.

Null Hypothesis A conservative way of phrasing the research question, to state that the independent variable has no effect on the dependent variable.

On-line Catalog A computerized reference index.

Open-ended Questions Respondents fill in their own answers.

Operational Definition Defining a variable by the means used to measure it. Intelligence can be operationally defined as a score on an IQ test.

Panel Study Frequently used in public opinion research, in which the same sample of respondents is interviewed on repeated occasions.

Participant Observation The observer becomes part of the event being studied and uses this position to make observations.

Pearson Product-moment Coefficient (r) A widely used statistic for measuring the relationship between two sets of scores that are assumed to be continuously distributed.

Peer Review A critical evaluation of a submitted paper by qualified people in the field.

Percentile The score at or below which a given percentage of cases lie.

Percentile Rank The percentage of cases in the comparison group at or below the score in question.

Person-centered Map A diagram showing people's movements and activities over a specified period of time.

Place-centered Map A diagram showing how people arrange themselves within a particular location.

Population The entire group of people, objects, or events in a category.

Predictive Validity The ability of a test to predict future behavior. A subtype of criterion validity.

Primary Source The original source of information.

Probability Level When used in statistics, it indicates the likelihood that an obtained difference on a statistical test is due to chance alone. A common practice in behavioral research is to use the .05 probability level (also called the .05 level of significance).

Probability Sample A sample that is drawn in such a way that the probability for the inclusion of any given individual can be estimated. Two general types of probability samples are *random samples* and *stratified samples.*

Probe A question or comment during an interview designed to keep the person talking or obtain clarification.

Program Evaluation Systematic procedures used to determine the effectiveness of a program.

Projective Test An instrument used in clinical testing and research that relies on questions or stimuli that are deliberately vague or incomplete.

Psychophysics The connection between the physical qualities of objects and their sensory attributes; very important in sensory evaluation.

Purposive Sample A type of nonprobability sample in which the individuals considered most relevant to the issue studied are selected for inclusion.

Quantification Expressing information in numbers.

Quasi-Experiment Also known as a natural experiment or ex post facto experiment. An experiment in which the independent variable occurs naturally and is not under the direct control of the experimenter.

Questionnaire A series of written questions on a topic about which the respondent's opinions are sought.

Quota Sample A type of nonprobability sampling in which quotas are assigned to characteristics according to the needs of the investigator, and then sampling is done to achieve these quotas.

r See *Pearson Product-moment Coefficient.*

Random Assignment Subjects in an experiment are randomly placed in treatment and control groups. A means of controlling extraneous variables.

Random Sample Every individual in the entire population being studied has an equal chance of being selected.

Range A measure of dispersion or variability, computed by subtracting the lowest from the highest score.

Raw Data Firsthand scores or measures. The actual measurements or data collected before they have been reduced, coded, or analyzed.

Reactivity The effect of the research upon the participants and the data collected.

Refereed Paper One that has been through a peer review process.

Regression Pertains to correlation; the relationships between sets of scores.

Reliability Consistency in measurement; the repeatability or replicability of findings.

Repeated Measures Design Research design involving the same subjects studied at different times or under different conditions.

Replication To repeat an earlier study to see if its findings can be duplicated. Important in increasing the confidence in research results.

Rho See *Spearman Rank-order Coefficient.*

Salience Refers to the importance of an issue in people's minds.

Sample The group of individuals selected for testing from a population is called a sample. The sample is used to make generalizations about the population from which it was drawn.

Sampling Bias The error introduced by a sampling procedure that favors certain characteristics over others.

Sampling Error Chance variation among samples selected from the same population.

Scale A series of ordered steps at fixed intervals used as a standard of measurement.

Secondary Source A second-hand report of information.

Semantic Differential A rating scale developed by Charles Osgood for measuring the meaning of concepts.

Simulation Deliberate creation of an artificial reality. Depends less on literal resemblance than on functional resemblance.

Skewed Distribution A distribution of scores that deviates markedly from a normal curve.

Snowball Sample A type of purposive (nonprobability) sample in which the researcher asks respondents for other people who should be contacted.

Software Programs for computers.

Spearman Rank-order Coefficient (rho) Correlation coefficient used with ranked data. Used in place of the Pearson product-moment correlation coefficient when a distribution contains extreme scores.

Split-Half Method A means of determining reliability by comparing two halves of a scale or test.

Standard Deviation A measure of dispersion or variability; sometimes abbreviated S.D. or σ. See Chapter 17 for method of computation.

Standardized Test Refers to a test that has been published with normative data and administered in a fixed way.

Step Scale On a questionnaire the respondent is required to select one of a graded series of intervals.

Stimulus–Response (S-R) Learning The development of associations between specific stimuli and specific responses. Theory developed by C. Hull.

Stratified Sample A type of probability sampling in which the characteristics of the sample are selected to be proportionate to those present in the total population.

Structured Interview Questions are formulated beforehand and asked in a set manner in a specified order. Often called a *standardized interview.*

Subjects Research participants or respondents.

Survey Research The systematic gathering of information about people's beliefs, attitudes, values, and behavior.

Systematic Observation Employs a scoring system and prearranged categories that are applied consistently.

t Test Statistical technique used for comparing two groups. See Chapter 18.

Test A systematic procedure for comparing people's performance, feelings, attitudes, or values.

Test Reliability The stability of measurement over time.

Test–Retest Method A means of determining reliability by administering a scale twice to the same people and comparing scores.

Thesaurus A dictionary of terms used as topic headings in an index such as Psychological Abstracts.

Thurstone Scale A type of attitude scale in which people check only those statements with which they agree. A respondent's score is computed by adding up the average rating of all statements checked, and dividing by the number of statements checked.

Time Sampling Procedure Specific times are selected according to a sampling plan for recording observed activities.

Trace Measures Physical remains of interaction after the individuals have departed. Two types of physical traces useful in behavioral research are accretion and erosion.

Transcription Putting interview or questionnaire responses into clear form for data analysis.

Treatment Condition Refers to the presence of the independent variable, as opposed to the control condition in which the independent (experimental) variable is absent.

Triangulation In observational research, refers to the use of multiple methods, observers, and sites to provide additional checks on a single observer's account.

Variability The amount of spread or dispersion within a distribution of scores. The more commonly used measures of variability are the range and the standard deviation.

Variable Any characteristic or quality that differs in degree or kind. Variables have values or levels and may either be continuous or categorical.

Value Label The label or name assigned to the variable level that is being represented as a number in a computer analysis (e.g., female = 1).

Unstructured Interview Although the interviewer has a general topic in mind, there is no predetermined order or specified wording to the questions.

Validity An indication that the measure accurately reflects what it is supposed to measure.

Value In research design, value refers to the amount of a variable (e.g., on a test of motor coordination with scores that can range from 1 to 20, the individual scores will represent values of the variable of motor coordination). Value has numerous other meanings in economics, philosophy, etc.

Variable Level The label or name representing a variable in a computer analysis, usually restricted to five or eight letters.

Vertical Axis The perpendicular line showing the graph boundary. Often used to indicate frequency.

Name Index

Subject Index